The Cambridge Star Atlas

THE CAMBRIDGE

STAR

ATLAS

SECOND EDITION

WIL TIRION

CAMBRIDGE
UNIVERSITY PRESS

Published by the Press Syndicate of the University of Cambridge
The Pitt Building, Trumpington Street, Cambridge CB2 1RP
40 West 20th Street, New York, NY 10011-4211, USA
10 Stamford Road, Oakleigh, Melbourne 3166, Australia

First published 1991
Second edition 1996

Printed in Great Britain at the University Press, Cambridge

A catalogue record for this book is available from the British Library

Library of Congress cataloguing in publication data

Tirion, Wil.
    Wil Tirion's Cambridge star atlas.
        p.   cm.
    Includes bibliographical references.
    ISBN 0–521–56098–5 (hc)
    1. Stars—Atlases.   I. Title.
QB65. T56   1996
523.8'0223—dc20   95–48882   CIP

ISBN 0 521 56098 5 hardback

CONTENTS

PREFACE

Anyone who looks up at the starry sky at night and wonders how to find a way among all those stars will need some kind of sky-guide or atlas, but very different needs must be met. The casual stargazer will first want to learn what can be seen with the unaided eye; the names of the stars, the constellations and where or when to look for Orion, the Great Bear, or Andromeda. The more advanced observer, with access to a good pair of binoculars or a small telescope, wants to know more: Where is the Whirlpool Galaxy, where the North-America Nebula, or the globular cluster M13?

The Cambridge Star Atlas offers help for both. It includes a series of twenty-four monthly sky maps, designed to be of use for almost anywhere on Earth together with a series of twenty detailed star charts, covering the whole heavens, with all stars visible to the naked eye under good circumstances. The star charts also show a wealth of star clusters, nebulae and galaxies. Some of these can be seen without optical help, but for most a small or average size telescope is needed. Accompanying the charts are tables, compiled by Patrick Moore, giving more information about the interesting objects on each chart.

Then there is a series of six all-sky maps, showing the sky in a special projection, with the Galactic Equator in the center, showing the distribution of stars, open and globular clusters, planetary and diffuse nebulae and galaxies, in relation to our own Milky Way.

Probably the most satisfying sight for a beginner, using a pair of binoculars or a small telescope, is the Moon. Therefore, in this second edition of the Cambridge Star Atlas, a not-too-complicated Moon map is added, showing the most important features on its surface. Since the Moon is our neighbour in space, and because it is usually the first thing we notice in the night sky, the Moon map is placed in front, where it belongs.

Happy stargazing!

Wil Tirion

THE MOON

The Moon is, apart from the Sun, the brightest object in the sky. Although the Sun and the Moon appear almost equal in size, they are quite different. The Sun is the central body of our Solar System, and all planets, including the Earth, orbit around it. The Sun measures 1.4 million kilometers across, and is at a distance of roughly 150 000 000 kilometers. The Moon is much smaller, 'only' 3476 kilometers across, approximately one quarter of the Earth's diameter, and at an average distance of 384 400 kilometers. It orbits not the Sun, but our own planet, in a little more than 27 days.

Although we often refer to the Moon as 'shining' it does not of itself give any light. It only reflects the light it receives from the Sun. This is the reason why the appearance of the Moon changes as it orbits the Earth. This aspect of the Moon, sometimes visible as a thin crescent in the western sky, after sunset, and sometimes as a full disk, lightening up the middle of the night, is confusing to most people. The reason for this can be best explained in a diagram (Figure 1).

The illustration is not drawn to scale, but shows you what happens. The Earth is at the center and the Moon's orbit is drawn as a circle. During its orbit around the Earth we see a different portion of the illuminated side of the Moon's surface. When the Moon is approximately between the Sun and the Earth, we see only the dark side of it. We call this *new Moon*. The Moon is not visible at all. After a few days we see a small crescent in the evening sky; a part of the illuminated side is peeping around the edge. After almost a week, half of the disk is lit and we call this *first quarter*. Another week later we see the complete disk. This is *full Moon*. Next comes the *last quarter*, and then back to *new Moon* again. From one new Moon to the next takes about 29.5 days, fully two days longer than it takes the Moon to orbit the Earth. The reason for this is that in the time the Moon revolves around the Earth, the Earth also moves in its orbit around the Sun.

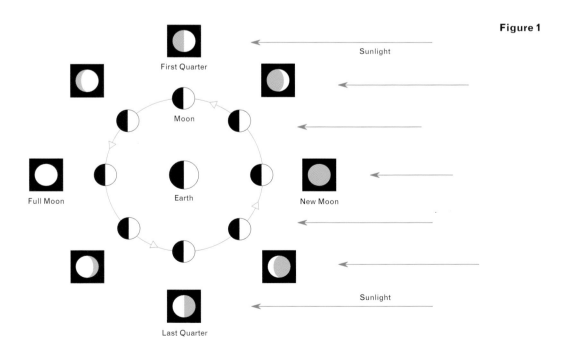

Figure 1

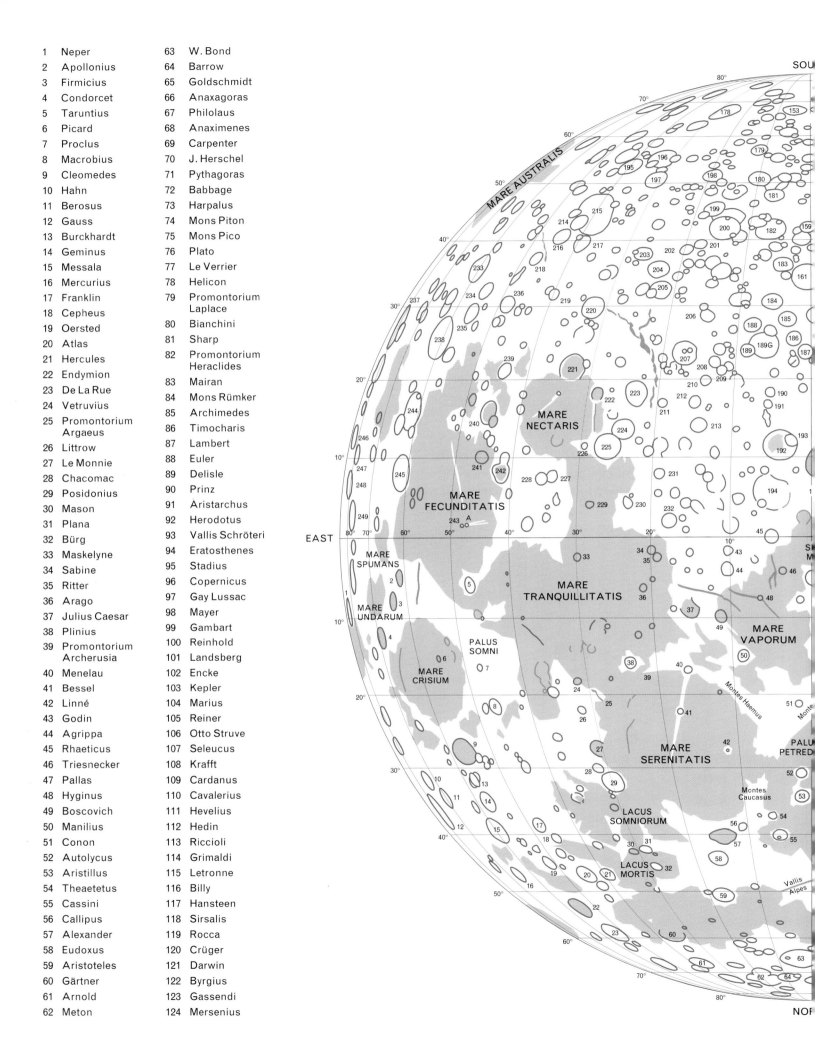

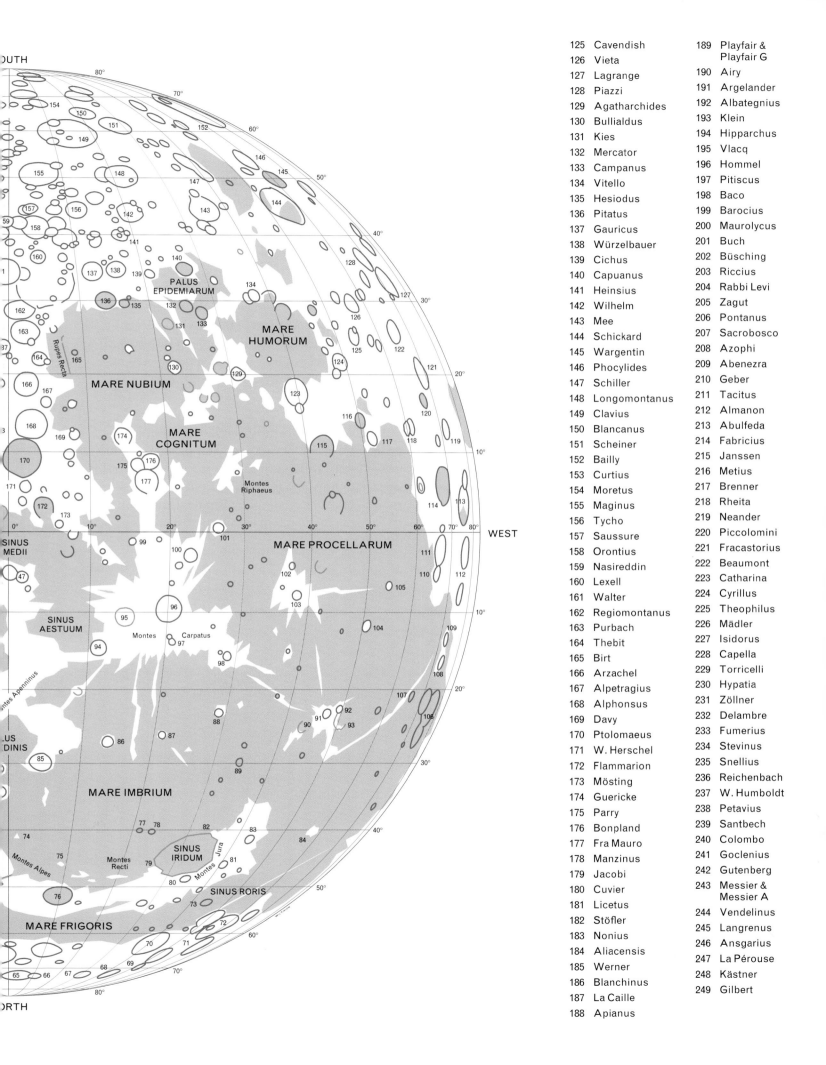

SOUTH

80°
70°
60°
50°
40°
30°
20°
10°

WEST

10°
20°
30°
40°
50°
60°
70°
80°

NORTH

PALUS
EPIDEMIARUM

MARE
HUMORUM

MARE NUBIUM

MARE
COGNITUM

Rupes Recta

Montes
Riphaeus

MARE PROCELLARUM

SINUS
MEDII

SINUS
AESTUUM

Montes Carpatus

MARE IMBRIUM

Montes Apenninus

PALUS
NUBINDIS

Montes Alpes

Montes Recti

SINUS
IRIDUM

Montes Jura

SINUS RORIS

MARE FRIGORIS

Even with a simple pair of binoculars you can see interesting features on the Moon's surface, and a small telescope will reveal even more interesting details on the surface of our neighbor in space. The best time to watch the Moon is not when it is full, but rather around its first or last quarter. Then the Moon is illuminated by the Sun from one side and especially near the terminator, the line dividing the lit and the unlit halves of the Moon, there is strong relief, because the surface is illuminated from a very low angle, resulting in long shadows. At full Moon you do not see any relief, since you are then looking from approximately the same direction as the Sun's rays come from. But full Moon is an ideal time to study the differences between the dark and light areas of the surface.

On the Moon map (on the previous two pages), all craters and crater-like features are drawn in dark green, while the larger *maria* are shown in a much lighter green. Maria is the plural form of the Latin word *mare*, meaning sea, and the name was given by the first observers who believed that these dark areas on the Moon really were seas and oceans. Although we now know there is not a drop of water on the Moon, the name persists, as also do the names *lacus* (lake) and *oceanus* (ocean). All *maria* and mountain ranges are labeled on the map itself, while the craters are numbered (to avoid overloading). You can find the names in the columns to the left and the right of the map.

Most craters on the Moon are believed to be the result of the impact of meteors: pieces of rock and metal from space. Our Earth is well protected against the impact of meteors by the atmosphere, which causes a meteor to burn and evaporate. Only the larger ones reach the surface; we call these meteorites. But the Moon does not have an atmosphere, so every meteor, captured by the Moon's gravity, will crash into the surface.

Because the Moon rotates 360° on its axis in exactly the same time that it takes to complete one orbit around the Earth, we always see the same side of the Moon. However the Moon's orbit is inclined about 5° to the *ecliptic* (see chapter *The star charts*, page 35), making it move slightly above and below the plane of the Earth's orbit around the Sun, and the Moon's axis is also tilted about 1.5°. The combined result is that we can also look about 6.5° 'over' the north and south pole. Moreover, since the Moon's orbit is not really a circle, but an ellipse, it does not move at a constant speed, though its rotation speed remains the same. Thus it moves a little from left to right, as if it were shaking its head very slowly. Therefore, sometimes we can look around the edges, by up to 7°. Sometimes *Mare Crisium* (in the northeastern quadrant of the Moon) appears very close to the edge, and sometimes it is closer to the center, and has a less elliptical appearance. The elliptical appearance of Mare Crisium, as well as that of craters close to the edge, is of course caused by perspective.

In most telescopes the image of the object that is viewed is upside down. For that reason the Moon map is placed with south on the top. So, if you are observing the Moon through a pair of binoculars, you will have to turn the book around. Unless you are living in the southern hemisphere; then it is the other way round!

THE MONTHLY SKY MAPS

Most people recognize one or two constellations; for example, they probably know what the Great Bear looks like. But why is it always in a different position in the sky? And why is it not possible to find Orion during a summer's night? This changing aspect of the sky is often confusing to the casual stargazer. So, the first thing one has to learn is how the sky moves.

It is important to know that the star-patterns themselves do not change, at least not in a single human life-span. It is only over a period of centuries that the positions of some neighboring stars change in a way that can be detected with the unaided eye. Now all these groupings of stars and constellations can be regarded as being fixed to a huge imaginary sphere, with the Earth placed in the center. No matter where on Earth we are, we can always see just one half of this sphere. So it is not hard to understand that, when we move to another part of the Earth, the visible part of the sphere also changes. If we stand at the North Pole we will only see the northern half of the heavens, while at the South Pole we will see only the southern half. But it is not that simple. Two more factors affect the appearance of the sky. First there is the daily rotation of the Earth around its axis, which causes the Sun to come up in the east and set in the west. The same thing happens with the other objects in the sky. In fact, the heavenly sphere seems to rotate around an axis that is an extension of the Earth's axis. Then there is the orbital movement of the Earth, which makes the appearance of the sky change over the seasons. If you look at one constellation, let us say Orion, at midnight on 1 January and take note of its position, and then look every successive night at about the same time, you will notice that the stars reach the same position a few minutes earlier each night. One month later, on 1 February, Orion will already be in that position at 10 PM. The appearance of the night sky thus changes slowly, until one year later the Earth has reached the same point in its orbit again and Orion will be back in its

original position at midnight. It is interesting to realize that if we were standing on the North Pole we would always see the same part of the sky, since the Earth's rotation will only cause us to turn around our own axis. The sky rotates around the point directly above us, the point in the sky we usually refer to as the zenith and stars only move parallel to the horizon, they do not rise or set. The celestial North Pole is at the zenith. At the South Pole we would see a similar situation, but with only the southern part of the sky visible. Since the axis of our planet does not change its position relative to the stars when it moves around the Sun the sky visible from the poles of the Earth remains the same all year long. In fact, of course, all six months long, the other six months will have daylight. On the other hand, on the Equator we will see a quite different situation. The celestial Equator, or the line where the plane of the Earth's Equator cuts the celestial sphere, runs from the east, passing directly overhead, to the west. All the stars and other objects in the sky rise and set, and during the year it is possible to view the entire celestial sphere. Finally, in the intermediate areas the situation is more complicated, as you can see in the illustration (Figure 2). There is an area of the sky that is called *circumpolar*. Stars in this part of the sky are so close to the pole that they never rise or set, but remain above the horizon. In the opposite part of the celestial sphere, there is an equal sized part that never becomes visible.

In the sky maps you can see a band of a slightly lighter blue, representing the brightest parts of the Milky Way. All the stars, nebulae and clusters we see in the sky (except the galaxies – see next chapter) belong to our own galaxy, the Milky Way. Since this galaxy is a huge flat disk-like formation we see most stars when we look along the plane of the Milky Way. All the light of the millions of stars that we do not see with the naked eye, forms a cloud-like band that is easily visible when you are outside the big cities and have bright skies.

The twenty-four monthly sky maps are constructed

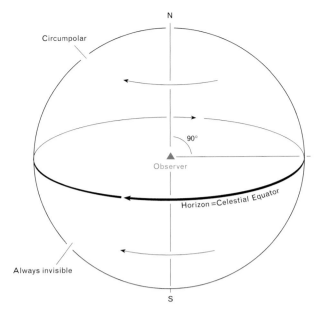

on a stereographic projection. Although this projection has the disadvantage that the scale increases from the center outward, its main advantage is that the shapes of the constellations and star groups are not distorted. When looking at the sky, there is a strange phenomenon, called the Moon illusion, which makes objects appear larger when close to the horizon than when they are high in the sky. This is, however, only a trick of the mind. We believe objects near the horizon to be at a greater distance than objects directly above our heads. The stereographic projection is in a way in harmony with this: constellations near the horizon appear larger!

Choosing the right map

There are twenty-four maps, two for each month, placed side-by-side on facing pages. The map on the left-hand page is for observers in the northern hemisphere and that on the right-hand page for southern observers. Each map has four different horizons, labeled to show you for what latitude on Earth the horizon is exact. Depending on the latitude on Earth at which you are living, you can select your horizon. A few degrees more or less will not make too much difference, when you are looking at the sky. The maps are for 11 PM on the first day of the given month, 10 PM on the 15th and 9 PM on the first day of the following month. These times are given at the bottom (left) of each map. But the maps can be used over a range of dates and times, as is shown in Table A. Let's say, you want to look at the sky in the middle of January, not in the evening, but at 6 AM. Simply look for January 15 in the left-hand column, and then look under 6 AM. You will find that you have to use the maps for the month of May. When local Summer Time (Daylight Saving Time; DST) is used, one hour should be added to the times given in the table (or one hour subtracted from the time on your watch).

The stars' magnitudes are given at the bottom right of each map. The word magnitude is explained in the introduction to the star charts, on page 36.

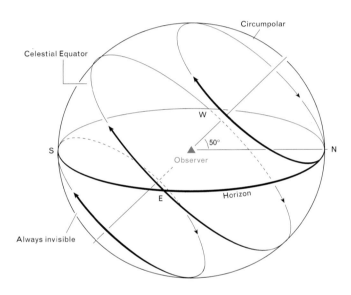

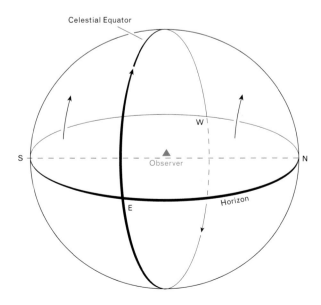

Figure 2

Table A *Selecting the monthly sky maps*

		5 PM	6 PM	7 PM	8 PM	9 PM	10 PM	11 PM	Midnight	1 AM	2 AM	3 AM	4 AM	5 AM	6 AM	7 AM
January	1	Oct		Nov		Dec		Jan		Feb		Mar		Apr		May
January	15		Nov		Dec		Jan		Feb		Mar		Apr		May	
February	1	Nov		Dec		Jan		Feb		Mar		Apr		May		Jun
February	15		Dec		Jan		Feb		Mar		Apr		May		Jun	
March	1	Dec		Jan		Feb		Mar		Apr		May		Jun		Jul
March	15		Jan		Feb		Mar		Apr		May		Jun		Jul	
April	1	Jan		Feb		Mar		Apr		May		Jun		Jul		Aug
April	15		Feb		Mar		Apr		May		Jun		Jul		Aug	
May	1	Feb		Mar		Apr		May		Jun		Jul		Aug		Sep
May	15		Mar		Apr		May		Jun		Jul		Aug		Sep	
June	1	Mar		Apr		May		Jun		Jul		Aug		Sep		Oct
June	15		Apr		May		Jun		Jul		Aug		Sep		Oct	
July	1	Apr		May		Jun		Jul		Aug		Sep		Oct		Nov
July	15		May		Jun		Jul		Aug		Sep		Oct		Nov	
August	1	May		Jun		Jul		Aug		Sep		Oct		Nov		Dec
August	15		Jun		Jul		Aug		Sep		Oct		Nov		Dec	
September	1	Jun		Jul		Aug		Sep		Oct		Nov		Dec		Jan
September	15		Jul		Aug		Sep		Oct		Nov		Dec		Jan	
October	1	Jul		Aug		Sep		Oct		Nov		Dec		Jan		Feb
October	15		Aug		Sep		Oct		Nov		Dec		Jan		Feb	
November	1	Aug		Sep		Oct		Nov		Dec		Jan		Feb		Mar
November	15		Sep		Oct		Nov		Dec		Jan		Feb		Mar	
December	1	Sep		Oct		Nov		Dec		Jan		Feb		Mar		Apr
December	15		Oct		Nov		Dec		Jan		Feb		Mar		Apr	

Northern latitudes

Date	Time	DST
January 1	11 pm	Midnight
January 15	10 pm	11 pm
February 1	9 pm	10 pm

Magnitudes:

●	●	●	●	·	·	·
−1	0	1	2	3	4	5

Southern latitudes

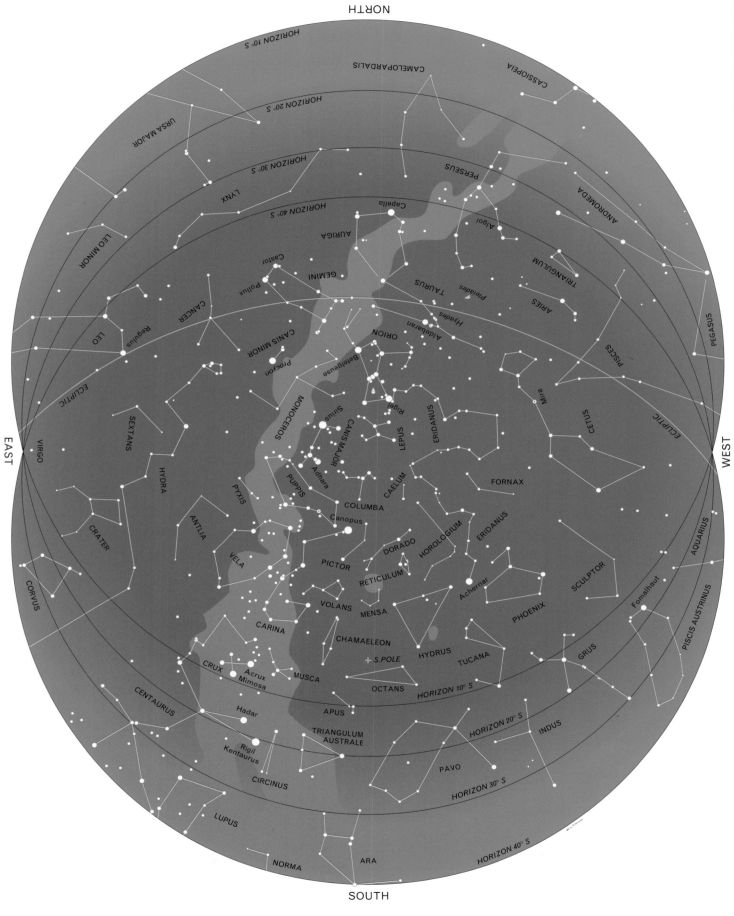

Date	Time	DST
January 1	11 pm	Midnight
January 15	10 pm	11 pm
February 1	9 pm	10 pm

Magnitudes:

−1 0 1 2 3 4 5

Northern latitudes

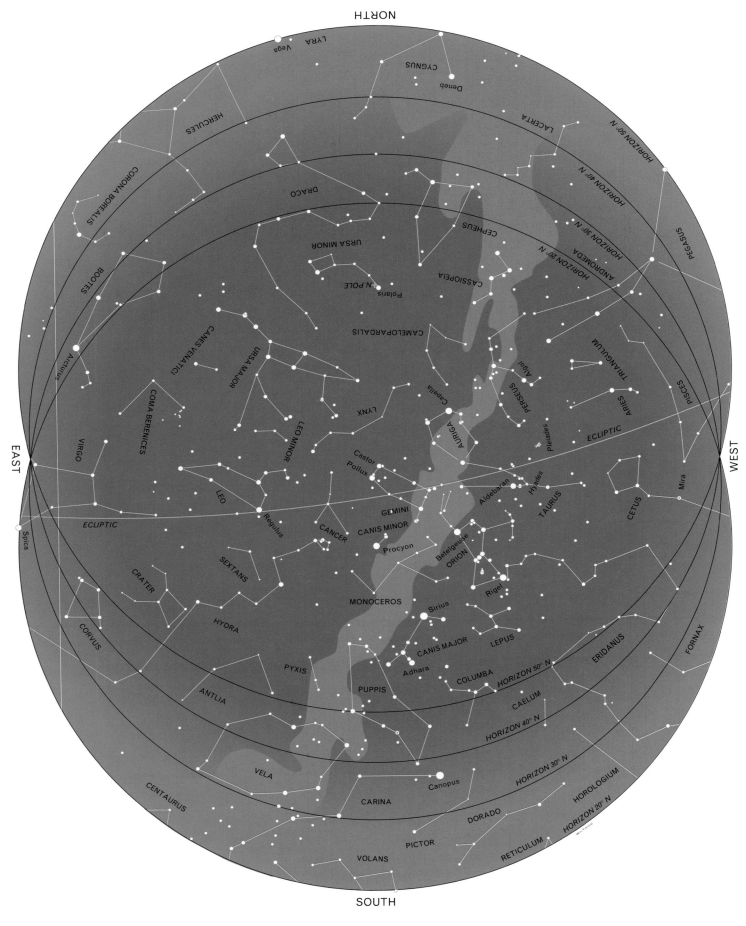

Date	Time	DST
February 1	11 pm	Midnight
February 15	10 pm	11 pm
March 1	9 pm	10 pm

Magnitudes:

−1	0	1	2	3	4	5

Southern latitudes

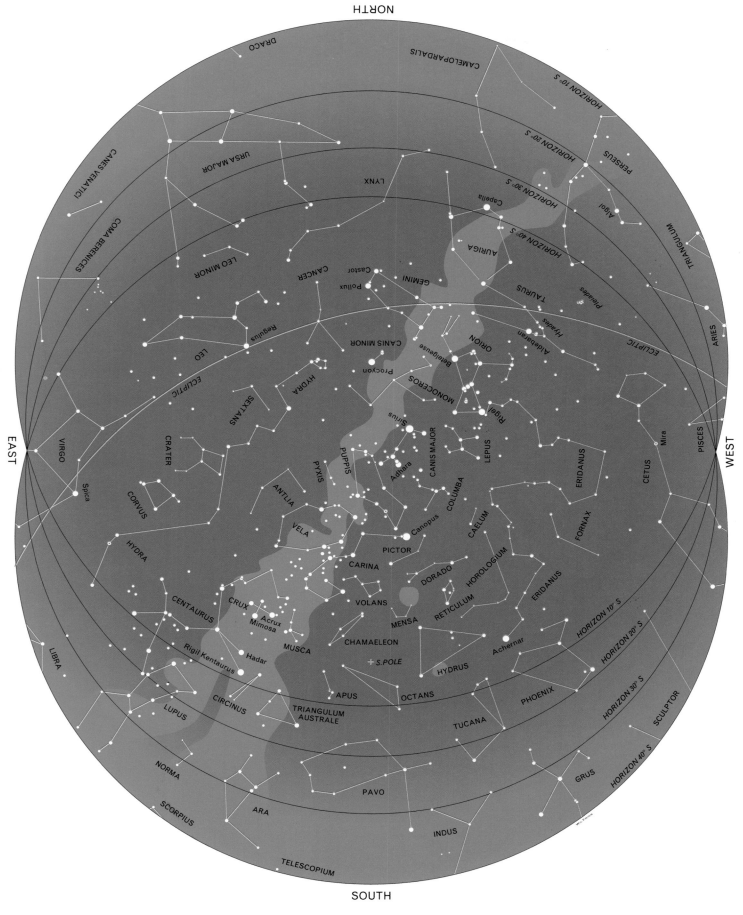

NORTH

DRACO

CAMELOPARDALIS

HORIZON 10° S

CANES VENATICI

URSA MAJOR

LYNX

PERSEUS

Algol

HORIZON 20° S

COMA BERENICES

LEO MINOR

HORIZON 30° S

Capella

HORIZON 40° S

TRIANGULUM

CANCER

Castor

GEMINI

AURIGA

TAURUS

Pleiades

LEO

Pollux

CANIS MINOR

Betelgeuse

ORION

Aldebaran

Hyades

ARIES

Regulus

ECLIPTIC

Procyon

MONOCEROS

Rigel

ECLIPTIC

EAST

SEXTANS

HYDRA

Sirius

CANIS MAJOR

LEPUS

Mira

PISCES

WEST

VIRGO

CRATER

Adhara

ERIDANUS

CETUS

Spica

PYXIS

PUPPIS

COLUMBA

FORNAX

CORVUS

ANTLIA

Canopus

CAELUM

HYDRA

VELA

PICTOR

ERIDANUS

CARINA

DORADO

HOROLOGIUM

HORIZON 10° S

CENTAURUS

CRUX

VOLANS

MENSA

RETICULUM

Achernar

HORIZON 20° S

Acrux
Mimosa

MUSCA

CHAMAELEON

HYDRUS

HORIZON 30° S

Rigil Kentaurus

Hadar

S. POLE

SCULPTOR

CIRCINUS

APUS

OCTANS

PHOENIX

LUPUS

TRIANGULUM
AUSTRALE

TUCANA

GRUS

HORIZON 40° S

NORMA

PAVO

INDUS

ARA

SCORPIUS

TELESCOPIUM

SOUTH

Date	Time	DST
February 1	11 pm	Midnight
February 15	10 pm	11 pm
March 1	9 pm	10 pm

Magnitudes:

−1	0	1	2	3	4	5

Northern latitudes

Date	Time	DST
March 1	11 pm	Midnight
March 15	10 pm	11 pm
April 1	9 pm	10 pm

Magnitudes:

−1 0 1 2 3 4 5

Southern latitudes

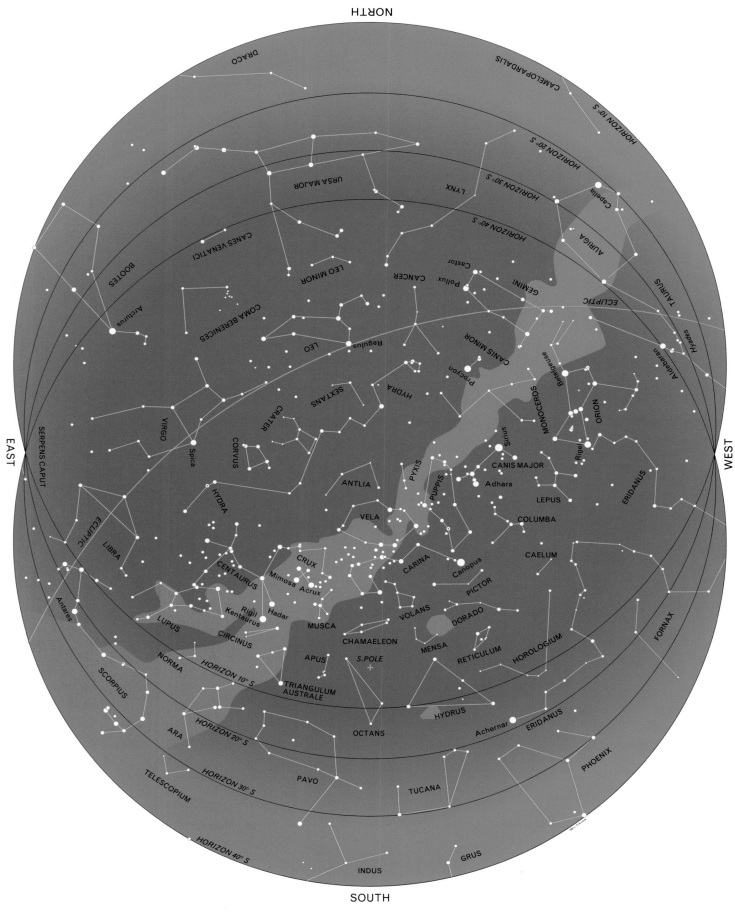

Date	Time	DST
March 1	11 pm	Midnight
March 15	10 pm	11 pm
April 1	9 pm	10 pm

Magnitudes:

−1	0	1	2	3	4	5

Northern latitudes

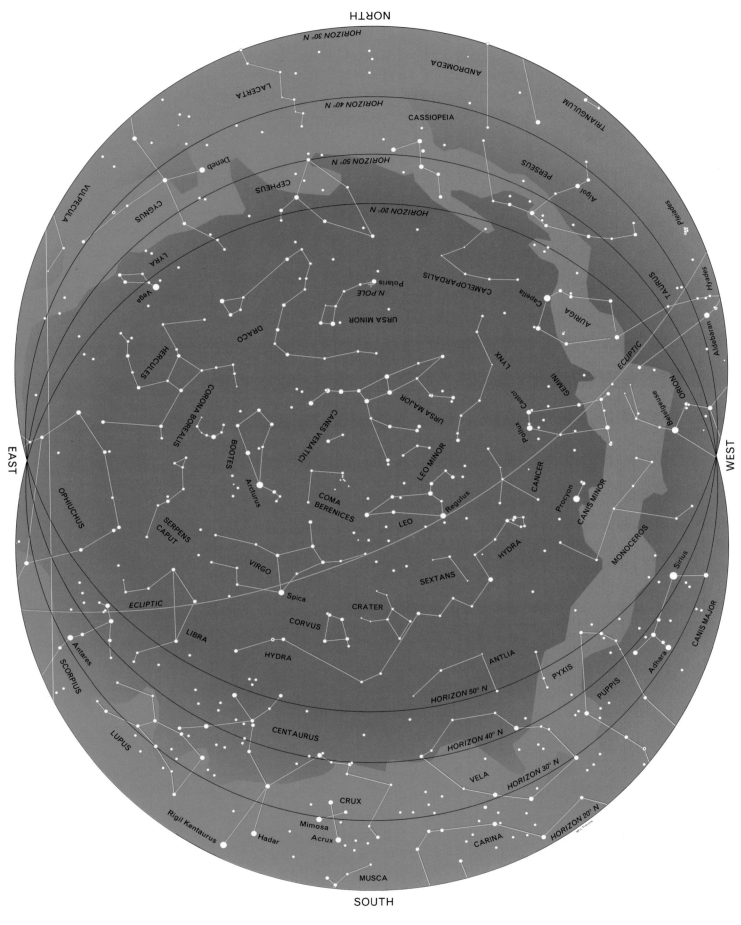

Date	Time	DST
April 1	11 pm	Midnight
April 15	10 pm	11 pm
May 1	9 pm	10 pm

Magnitudes:

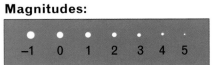

−1 0 1 2 3 4 5

Southern latitudes

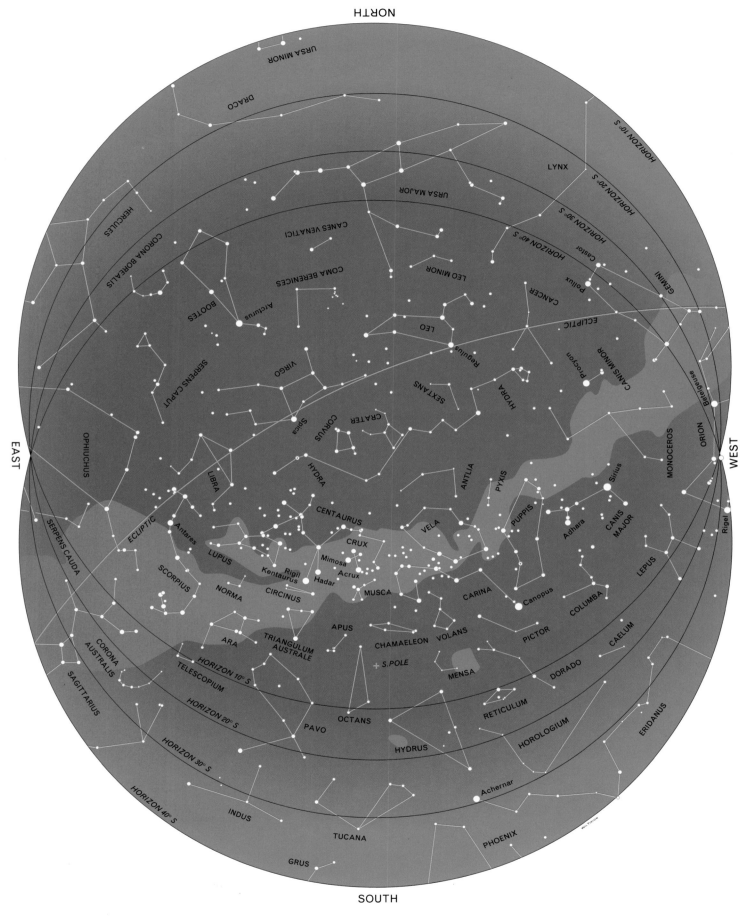

Date	Time	DST
April 1	11 pm	Midnight
April 15	10 pm	11 pm
May 1	9 pm	10 pm

Magnitudes:

−1   0   1   2   3   4   5

Northern latitudes

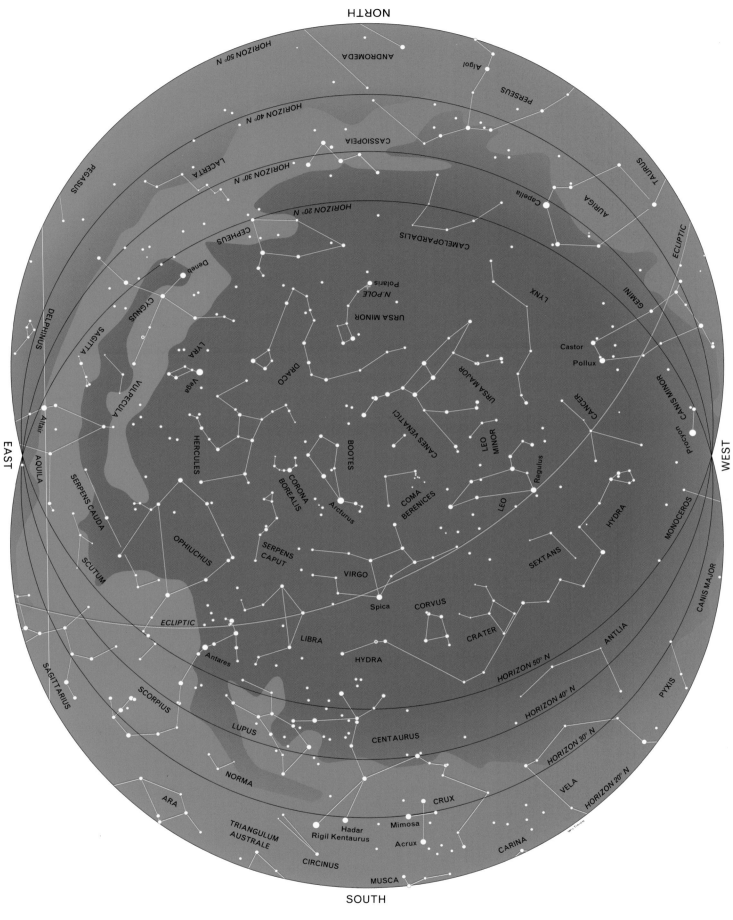

Date	Time	DST
May 1	11 pm	Midnight
May 15	10 pm	11 pm
June 1	9 pm	10 pm

Magnitudes:

−1 0 1 2 3 4 5

Southern latitudes

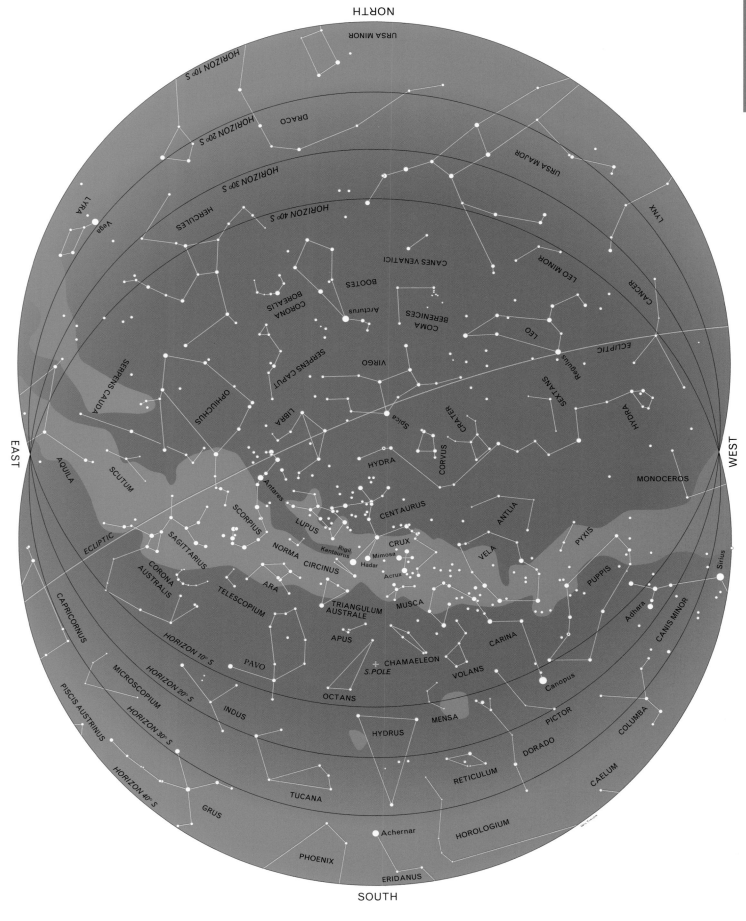

NORTH

URSA MINOR

HORIZON 10° S

DRACO

HORIZON 20° S

URSA MAJOR

HORIZON 30° S

HERCULES

HORIZON 40° S

LYRA

Vega

LYNX

CANES VENATICI

BOOTES

CORONA BOREALIS

Arcturus

COMA BERENICES

LEO MINOR

CANCER

SERPENS CAPUT

OPHIUCHUS

VIRGO

LEO

Regulus

ECLIPTIC

SERPENS CAUDA

LIBRA

Spica

CRATER

SEXTANS

HYDRA

EAST

WEST

AQUILA

SCUTUM

HYDRA

CORVUS

MONOCEROS

Antares

CENTAURUS

SCORPIUS

LUPUS

ANTLIA

ECLIPTIC

SAGITTARIUS

NORMA

Rigil
Kentaurus

CRUX

VELA

CORONA
AUSTRALIS

CIRCINUS

Mimosa

Hadar

Acrux

PYXIS

PUPPIS

Sirius

CAPRICORNUS

TELESCOPIUM

ARA

MUSCA

CARINA

Adhara

CANIS MINOR

TRIANGULUM
AUSTRALE

APUS

CHAMAELEON

VOLANS

CANOPUS

PICTOR

COLUMBA

MICROSCOPIUM

S. POLE

CARINA

HORIZON 10° S

PAVO

Canopus

PISCIS AUSTRINUS

HORIZON 20° S

INDUS

OCTANS

MENSA

PICTOR

CAELUM

HORIZON 30° S

HYDRUS

DORADO

HORIZON 40° S

TUCANA

RETICULUM

HOROLOGIUM

GRUS

Achernar

PHOENIX

ERIDANUS

SOUTH

Date	Time	DST
May 1	11 pm	Midnight
May 15	10 pm	11 pm
June 1	9 pm	10 pm

Magnitudes:

−1	0	1	2	3	4	5

May

Northern latitudes

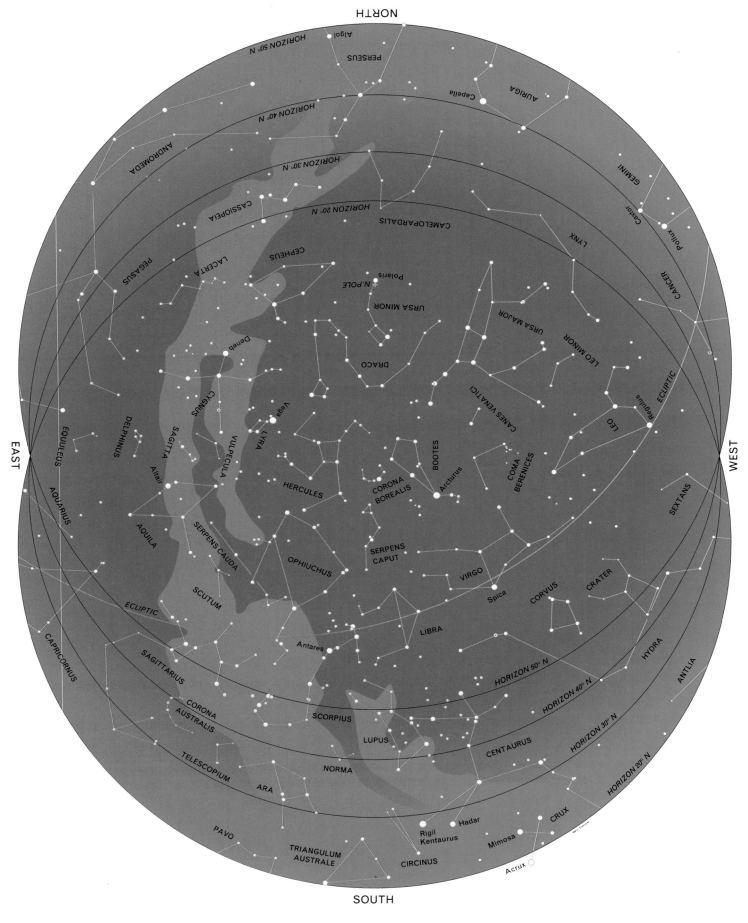

Date	Time	DST
June 1	11 pm	Midnight
June 15	10 pm	11 pm
July 1	9 pm	10 pm

Magnitudes:

−1 0 1 2 3 4 5

Southern latitudes

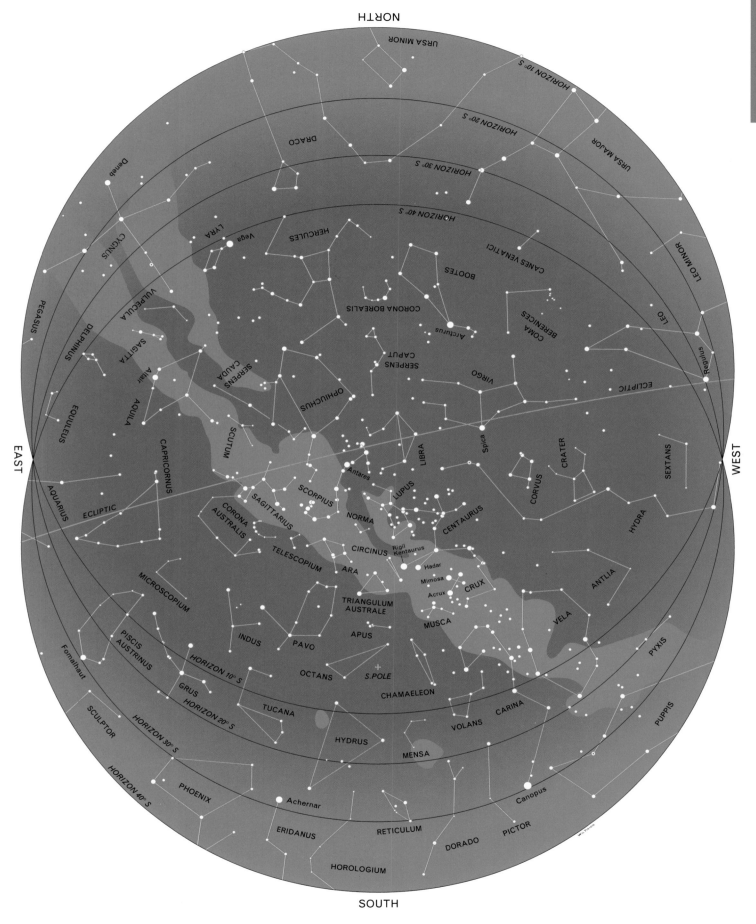

Date	Time	DST
June 1	11 pm	Midnight
June 15	10 pm	11 pm
July 1	9 pm	10 pm

Magnitudes:

−1 0 1 2 3 4 5

Northern latitudes

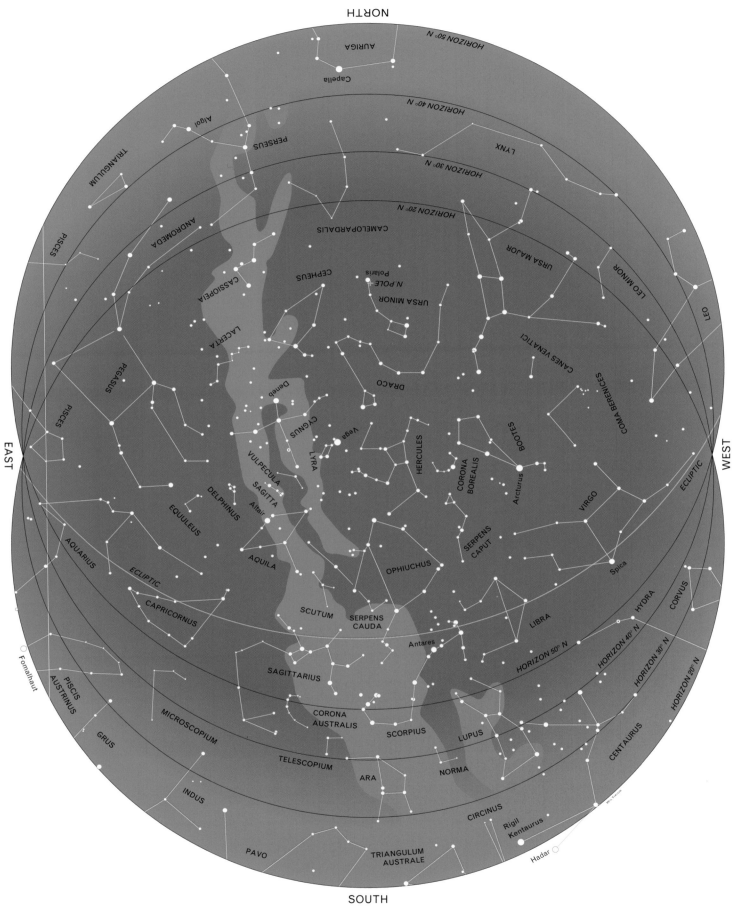

NORTH

HORIZON 50° N
HORIZON 40° N
HORIZON 30° N
HORIZON 20° N

AURIGA
Capella
Algol
PERSEUS
TRIANGULUM
ANDROMEDA
PISCES
PISCES
PEGASUS
LACERTA
CASSIOPEIA
CEPHEUS
CAMELOPARDALIS
LYNX
URSA MAJOR
LEO MINOR
LEO
Polaris
N.POLE
URSA MINOR
DRACO
CANES VENATICI
COMA BERENICES
Deneb
CYGNUS
VULPECULA
SAGITTA
LYRA
Vega
HERCULES
BOOTES
Arcturus
EAST
DELPHINUS
Altair
EQUULEUS
AQUILA
CORONA
BOREALIS
SERPENS
CAPUT
VIRGO
WEST
ECLIPTIC
AQUARIUS
ECLIPTIC
SCUTUM
OPHIUCHUS
Spica
CAPRICORNUS
SERPENS
CAUDA
LIBRA
HYDRA
CORVUS
Antares
HORIZON 50° N
HORIZON 40° N
HORIZON 30° N
HORIZON 20° N
Fomalhaut
PISCIS
AUSTRINUS
MICROSCOPIUM
SAGITTARIUS
CORONA
AUSTRALIS
SCORPIUS
LUPUS
CENTAURUS
GRUS
TELESCOPIUM
ARA
NORMA
INDUS
CIRCINUS
Rigil
Kentarus
PAVO
TRIANGULUM
AUSTRALE
Hadar

SOUTH

Date	Time	DST
July 1	11 pm	Midnight
July 15	10 pm	11 pm
August 1	9 pm	10 pm

Magnitudes:

−1 0 1 2 3 4 5

Southern latitudes

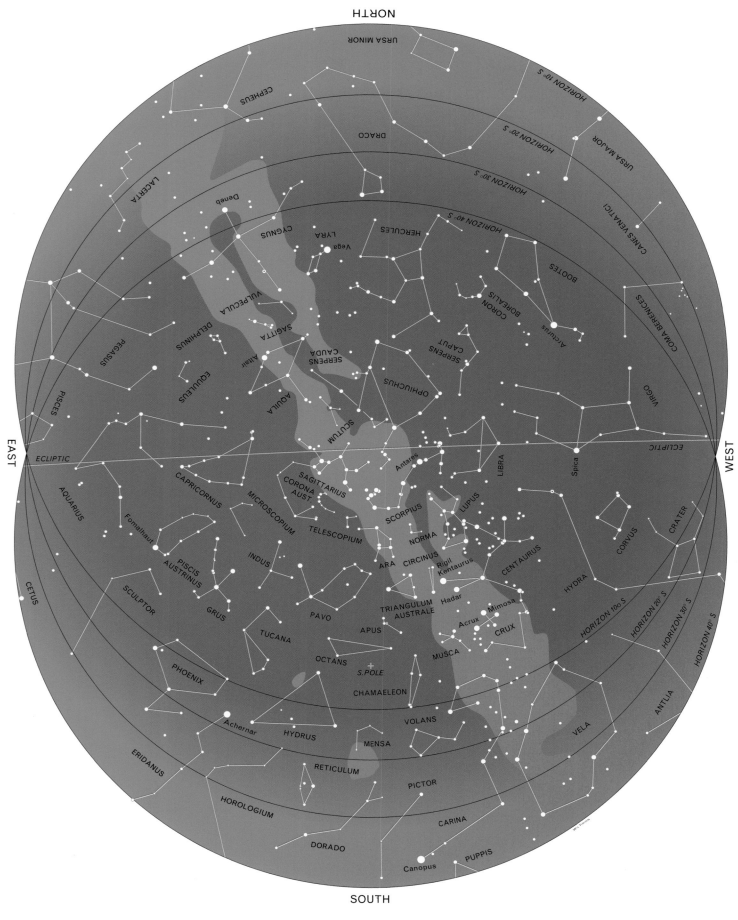

NORTH

URSA MINOR

CEPHEUS

DRACO

LACERTA

HORIZON 10° S

URSA MAJOR

HORIZON 20° S

CANES VENATICI

HORIZON 30° S

HERCULES

HORIZON 40° S

Deneb

CYGNUS

LYRA
Vega

BOOTES

VULPECULA

CORONA BOREALIS

SAGITTA
Altair

SERPENS CAPUT

Arcturus

PEGASUS

DELPHINUS

SERPENS CAUDA

COMA BERENICES

PISCES

EQUULEUS

AQUILA

OPHIUCHUS

VIRGO

EAST

ECLIPTIC

SCUTUM

ECLIPTIC

WEST

AQUARIUS

CAPRICORNUS

MICROSCOPIUM

SAGITTARIUS
CORONA AUST.

Antares

LIBRA

Spica

Fomalhaut

SCORPIUS

LUPUS

CRATER

PISCIS AUSTRINUS

INDUS

TELESCOPIUM

NORMA

CORVUS

CETUS

GRUS

PAVO

ARA CIRCINUS

Rigil Kentaurus

CENTAURUS

HYDRA

SCULPTOR

TUCANA

TRIANGULUM AUSTRALE

Hadar

Mimosa

HORIZON 10° S

HORIZON 20° S

PHOENIX

APUS

Acrux

CRUX

HORIZON 30° S

ANTLIA

OCTANS

MUSCA

HORIZON 40° S

Achernar

HYDRUS

S. POLE

CHAMAELEON

ERIDANUS

MENSA

VOLANS

VELA

RETICULUM

PICTOR

HOROLOGIUM

CARINA

DORADO

Canopus

PUPPIS

SOUTH

Date	Time	DST
July 1	11 pm	Midnight
July 15	10 pm	11 pm
August 1	9 pm	10 pm

Magnitudes:

−1 0 1 2 3 4 5

Northern latitudes

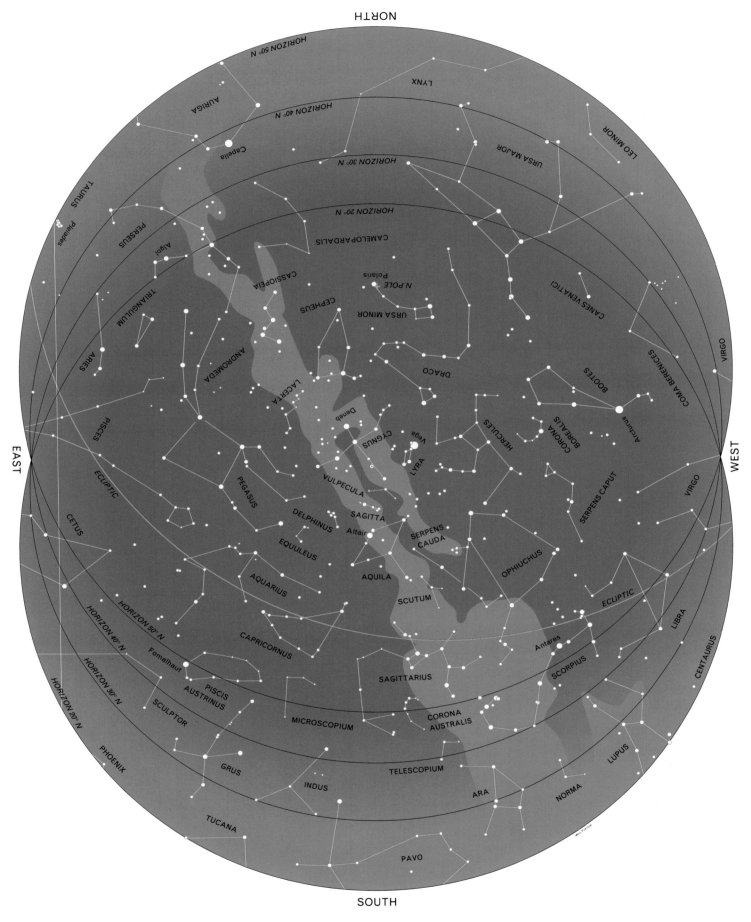

Date	Time	DST
August 1	11 pm	Midnight
August 15	10 pm	11 pm
September 1	9 pm	10 pm

Magnitudes:

−1 0 1 2 3 4 5

Southern latitudes

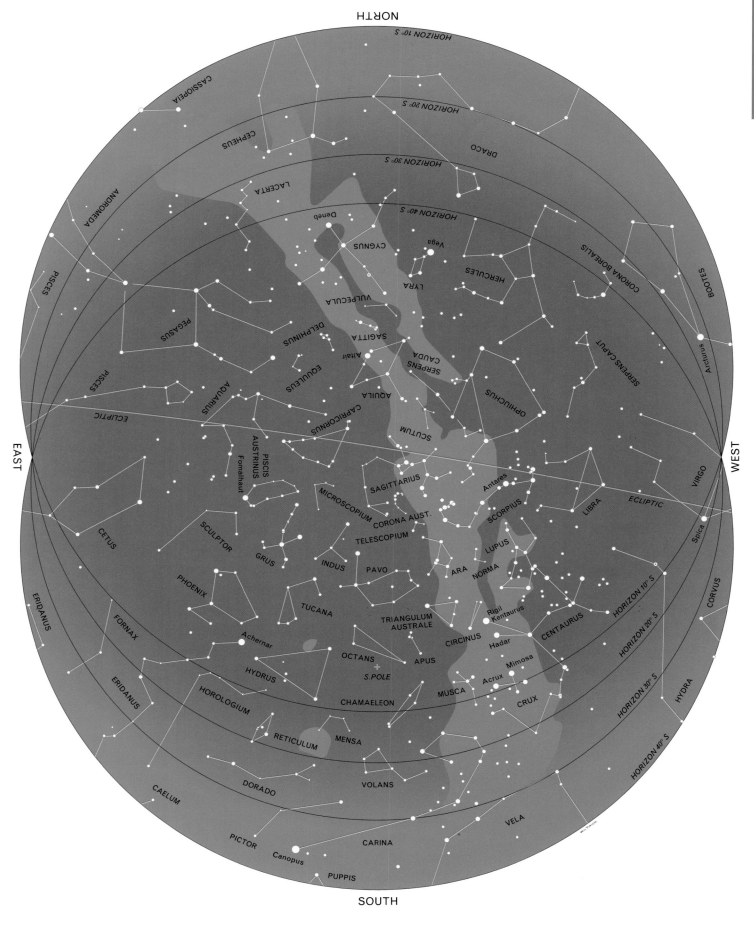

Date	Time	DST
August 1	11 pm	Midnight
August 15	10 pm	11 pm
September 1	9 pm	10 pm

Magnitudes:

−1 0 1 2 3 4 5

Northern latitudes

Date	Time	DST
September 1	11 pm	Midnight
September 15	10 pm	11 pm
October 1	9 pm	10 pm

Magnitudes:

−1 0 1 2 3 4 5

Southern latitudes

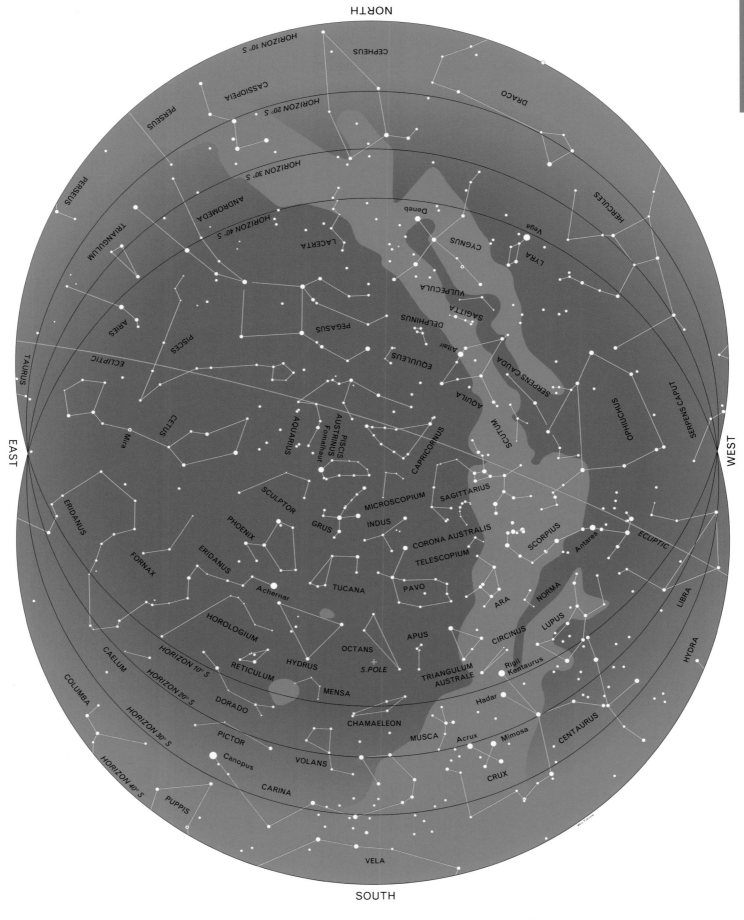

Date	Time	DST
September 1	11 pm	Midnight
September 15	10 pm	11 pm
October 1	9 pm	10 pm

Magnitudes:

−1 0 1 2 3 4 5

Northern latitudes

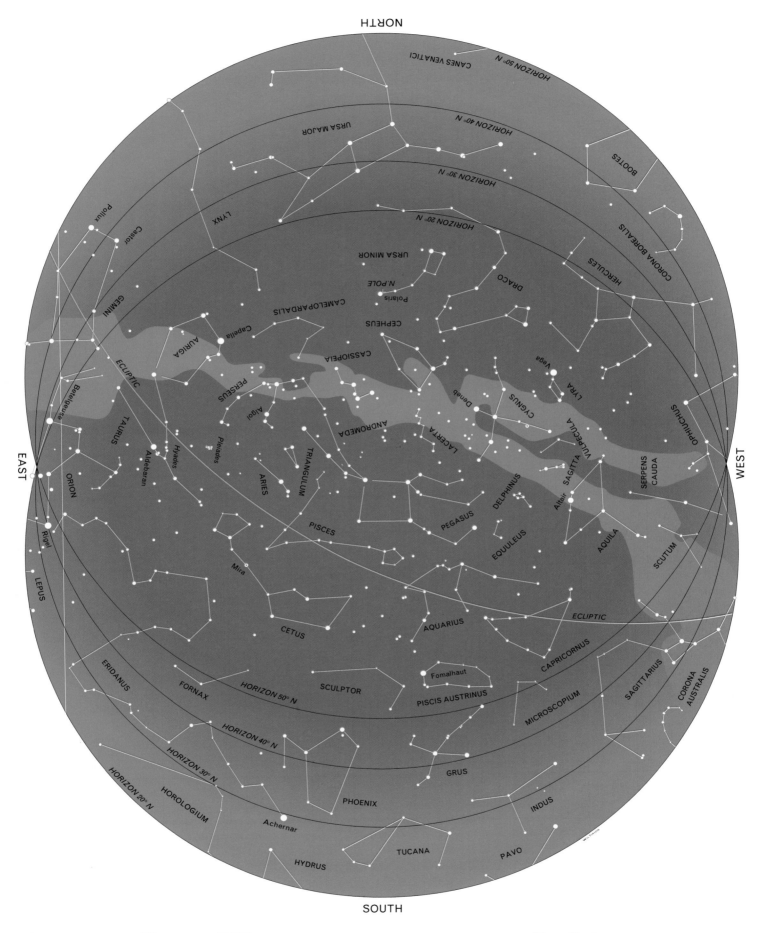

NORTH

SOUTH

EAST

WEST

Date	Time	DST
October 1	11 pm	Midnight
October 15	10 pm	11 pm
November 1	9 pm	10 pm

Magnitudes:

−1 0 1 2 3 4 5

Southern latitudes

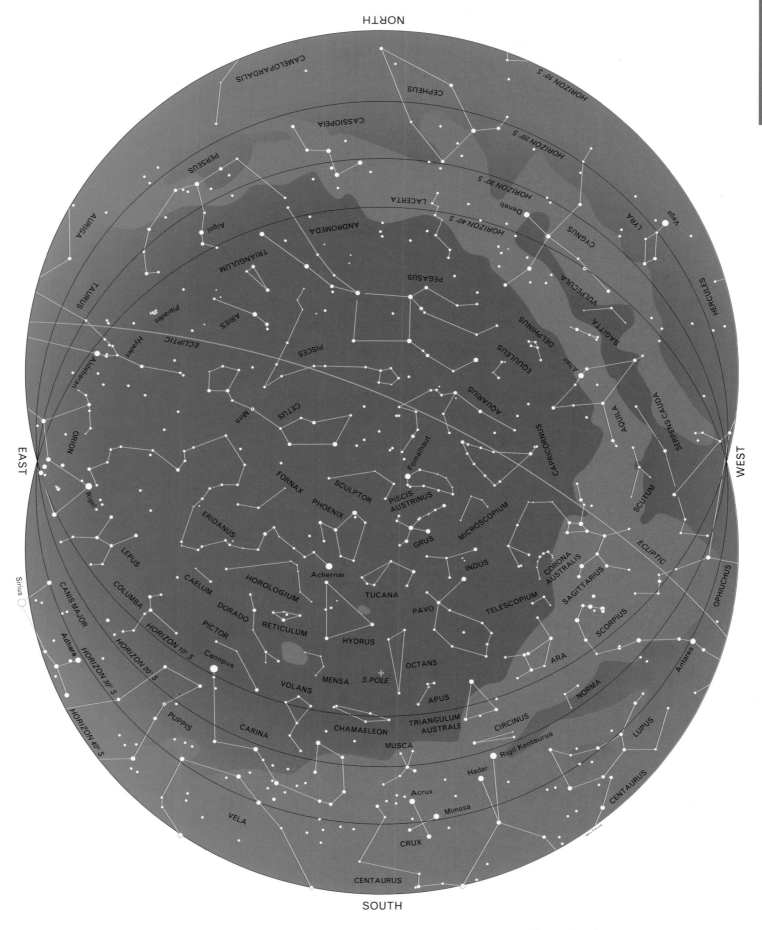

Date	Time	DST
October 1	11 pm	Midnight
October 15	10 pm	11 pm
November 1	9 pm	10 pm

Magnitudes:

−1 0 1 2 3 4 5

Northern latitudes

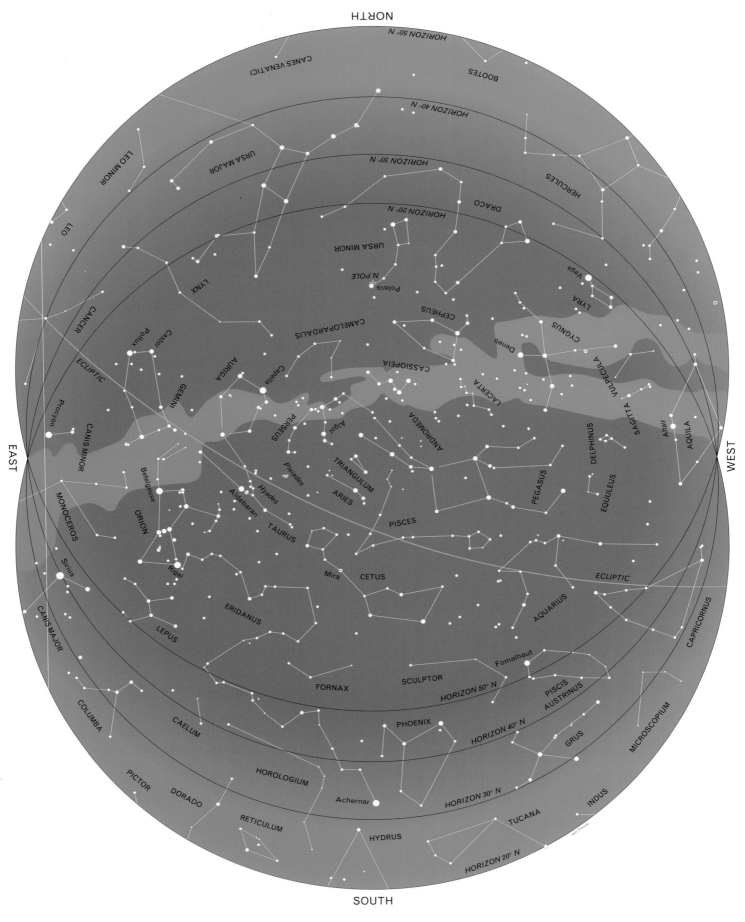

Date	Time	DST
November 1	11 pm	Midnight
November 15	10 pm	11 pm
December 1	9 pm	10 pm

Magnitudes:

−1　0　1　2　3　4　5

Southern latitudes

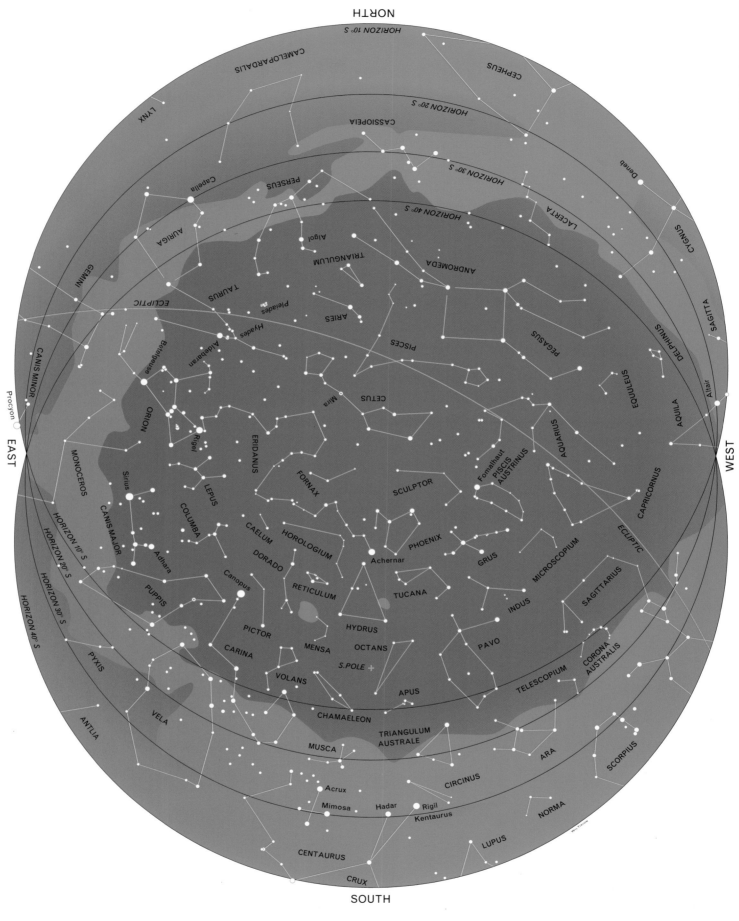

Date	Time	DST
November 1	11 pm	Midnight
November 15	10 pm	11 pm
December 1	9 pm	10 pm

Magnitudes:

−1 0 1 2 3 4 5

Northern latitudes

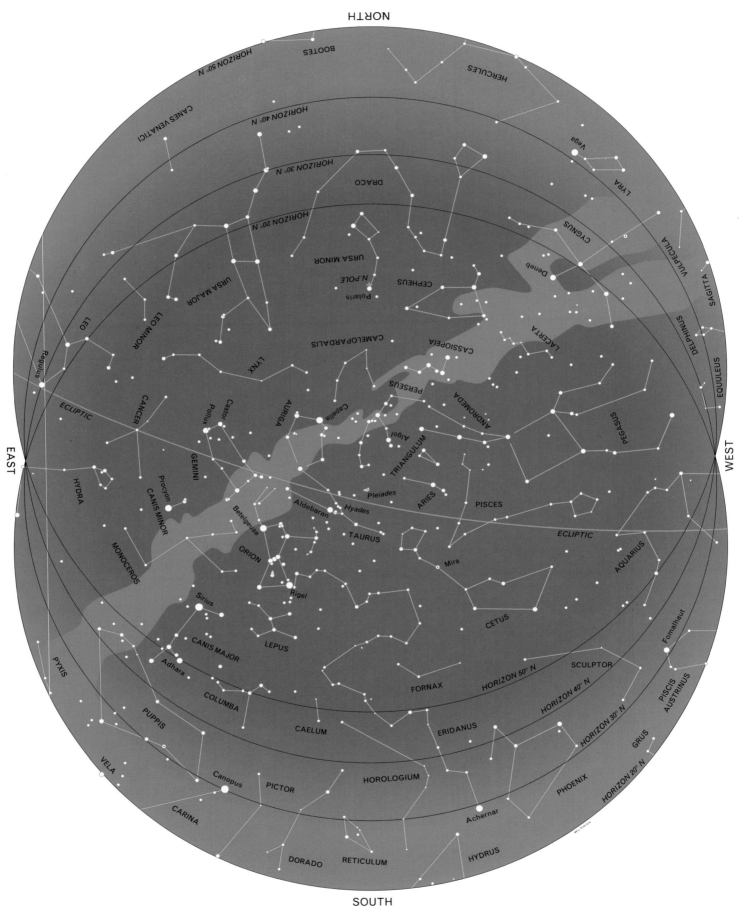

Date	Time	DST
December 1	11 pm	Midnight
December 15	10 pm	11 pm
January 1	9 pm	10 pm

Magnitudes:

−1 0 1 2 3 4 5

Southern latitudes

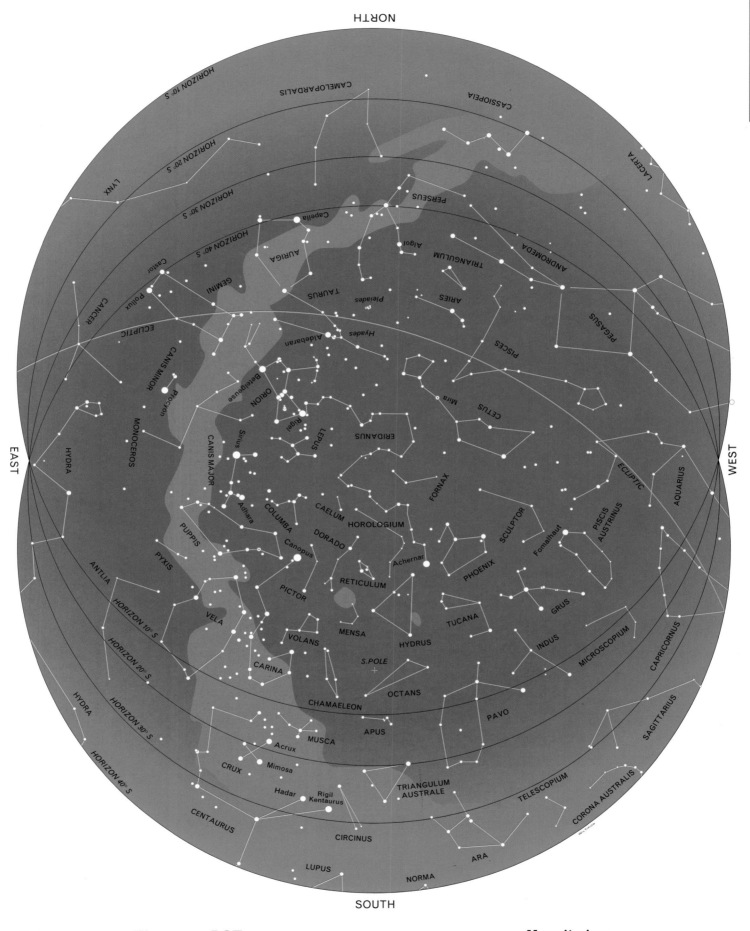

Date	Time	DST
December 1	11 pm	Midnight
December 15	10 pm	11 pm
January 1	9 pm	10 pm

Magnitudes:

−1 0 1 2 3 4 5

THE STAR CHARTS

Once you are familiar with the sky and can recognize the different constellations, you will want to know more about what is visible in the night sky. So, the next step is to use the star charts in this chapter, the 'heart' of the *Cambridge Star Atlas*. These star charts divide the sky into 20 parts. The actual chart areas are shown on the chart index on page 43, preceding the star charts themselves. There is a generous overlap between the charts, so most of the constellations are shown complete on at least one chart. The positions of stars and objects are for the year 2000, or to be more precise, the epoch is 2000.0, the extra 0 is a decimal and it means 1 January. (2000.5 would be 1 June.) The positions are plotted against a grid of right ascension (RA) and declination (Dec) comparable with longitude and latitude on the Earth's globe. *Right ascension* is reckoned in hours, minutes and seconds from 0h to 24h, from west to east along the Equator. *Declination* represents the angular distance between an object and the celestial Equator, (+) for objects north and (–) for those south of the Equator (Figure 3).

The charts' projections have been carefully chosen to show the star-patterns with the least possible distortion and to make it easy to measure positions from the maps. The projections used are *azimuthal equidistant* (the polar charts, 1 and 20), *secant cylindrical* (the equatorial band, charts 8–13) and the *secant conic projection* (the intermediate areas). The advantage of all three projections is that the hour circles (lines of equal RA; running north/south from pole to pole, like the meridians on Earth) are shown as straight lines, and all parallels (lines of equal declination, parallel to the Equator) are equally spaced everywhere on the charts, so positions can easily be measured with a simple ruler. The tick marks along the chart borders and on the central hour circle on the conic maps will be helpful.

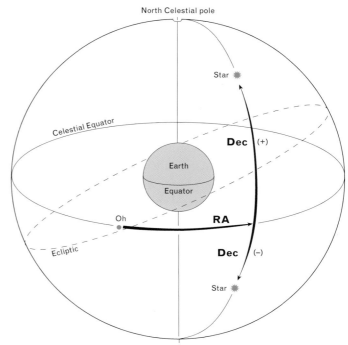

Figure 3

Magnitudes

The word magnitude usually refers to the apparent brightness of a star or an object. Traditionally, stars visible to the naked eye were binned into six groups of brightnesses. The most prominent stars in the sky were called first magnitude stars, the ones slightly fainter (e.g. the Pole Star) second magnitude and so on. The faintest ones visible to the eye were magnitude six. Nowadays astronomers are able to measure the brightness very accurately. In modern catalogues you will find magnitude to two places of decimals and the scale now has a logarithmic footing. A difference of five magnitudes is defined as 100. Consequently one magnitude represents a difference of 2.5 times (or to be precise 2.512, the fifth root of 100). On this scale several stars turned out to be brighter than 1, so the scale was extended to 0, but even that was not enough. A few stars were still brighter and were given a negative magnitude. The brightest star in the sky, Sirius, has a magnitude of −1.46.

In the atlas, as well as on the twenty-four monthly sky maps, the brightness of a star is rounded to the nearest whole magnitude. Stars between 0.51 and 1.50 are binned to magnitude 1, between 1.51 and 2.50 magnitude 2 and so on. Positions and magnitudes are taken from the *Bright Star Catalogue* by Dorrit Hoffleit (fourth edition, 1982). All 9096 stars from the catalogue are plotted. Stars having a magnitude of 6.5 or brighter in the *Smithsonian Astrophysical Observatory Star Catalog* and not included in the *Bright Star Catalogue* have also been added, bringing the total of plotted stars to ±9500. Furthermore, the atlas shows 866 non-stellar objects, i.e. star clusters, nebulae and distant galaxies.

Stars are huge balls of hot gas, like our own Sun, but much further away than the Sun. Although they appear as small points of light (so small that even in the largest telescope most of them cannot be measured) many stars are even bigger than our Sun, which turns out to be of only average size. Some stars, like Betelgeuse (in Orion) and Antares (in Scorpius) have diameters that are close to half a billion kilometers (300 times that of our Sun). Even that is not the limit: some stars have diameters a thousand times larger than our Sun. At the other end of the scale are stars that are no bigger than the Earth, and some that are even smaller.

Not only the sizes, but also the temperatures of stars differ. The temperature of the Sun is almost 5500 °C. That is the surface temperature; in the nucleus of the Sun the temperature reaches millions of degrees. The Sun is a star that is radiating yellowish light. 'Cooler' stars are orange (like Aldebaran in Taurus, or Arcturus in Boötes, with surface temperatures of 4000 °C) or red (Betelgeuse and Antares; 3000 °C). But many stars are much hotter than the Sun. White stars have temperatures between 6000 and 10 000 °C (Sirius, in Canis Major, or Deneb in Cygnus). Then there are the bluish stars with temperatures between 11 000 and 40 000 °C (like Rigel in Orion). It is interesting to take a look at the beautiful constellation Orion on a bright winter's night. You can clearly see the difference between the red giant star Betelgeuse and the bluish Rigel. Try it.

Stars are divided into spectral classes, according to differences in their spectra. These differences are related to the surface temperatures and the colors of the stars. The hottest, bluish stars are the classes O, B and A, the slightly cooler stars, white, are class F, yellow class G, orange class K and the coolest, the red stars are class M. To remember this rather illogical sequence of letters, remember the sentence: **Oh, Be A Fine Girl, Kiss Me!**, or if you prefer: **Oh, Be A Fine Guy, Kiss Me!** In the tables, facing the star charts, the spectral classes of variable stars are given in the final column.

Star names

Many of the brighter stars have proper names of Latin, Greek or Arabic origin, such as Regulus, Altair and Betelgeuse. Only the brightest stars are still referred to by these old names. Most stars now bear designations of numbers and Greek letters. The German celestial cartographer Johann Bayer introduced the Greek letters in the beginning of the seventeenth century. In general, the brightest star in a constellation was given the first letter of the Greek alphabet, alpha (α), the second letter, beta (β), was given to the second brightest and so on, although there are many obvious exceptions to this rule. In Table B you will find the Greek letters and their names.

Another way to identify stars is by numbers. In each constellation the stars are numbered in order of RA. These numbers are usually referred to as Flamsteed numbers. Most of the brighter stars have both a Greek letter and a number. On the star charts you will find all Greek letters and in addition, the Flamsteed numbers for those stars not having a Greek letter. For stars of magnitude 1 and brighter, the proper names are also given, as is done for a few well-known second magnitude stars: Algol, Mira, Castor and Polaris.

Whenever a star is referred to by a Greek letter or by a number, this is followed by the genitive form of the Latin constellation name, or its official abbreviation. So the star Deneb, in the Swan (Latin: *Cygnus*) can also be called Alpha (α) Cygni or 50 Cygni (α Cyg, 50 Cyg). Table C gives the names of the constellations, the genitive form, the official abbreviation and the common English names. Variable stars have a quite different nomenclature. Some have the regular star identification like Algol (β Per) or Mira (o Cet) but most are labeled in a special way: by roman letters starting with R, then S, T etc. to Z. Then RR, RS, RT, …, RZ, next SS, ST, and so on up to ZZ. After these 54 the naming continues with AA, AB, AC, … , AZ, then BB, BC and so on again up to QZ, giving a total of 334. The next variable found in that constellation is called V335, then V336 and so on. These designations are also followed by the constellation name, as with the Greek letters and the Flamsteed numbers.

Table B *The Greek alphabet*

α	Alpha	ν	Nu
β	Beta	ξ	Xi
γ	Gamma	o	Omicron
δ	Delta	π	Pi
ε	Epsilon	ρ	Rho
ζ	Zeta	σ	Sigma
η	Eta	τ	Tau
θ	Theta	υ	Upsilon
ι	Iota	ϕ	Phi
κ	Kappa	χ	Chi
λ	Lambda	ψ	Psi
μ	Mu	ω	Omega

Constellations

Most of the constellations we know originate from Mesopotamian traditions and from Greek mythology, but over the centuries many other constellations have been added to the classical ones, especially in the southern sky. Several of these new constellations only had a short life, while others have survived. In 1930 the International Astronomical Union finally adopted a list of 88 official constellations and the boundaries were also delimited once and for all. On our charts these official boundaries are drawn in as broken lines. The final column of Table C gives the charts where you can find each of these constellations.

Table C *List of constellations*

Name	Genitive	Abbreviation	Common name	Chart number(s)		
Andromeda	Andromedae	And	*Andromeda*	2		
Antlia	Antliae	Ant	*Air Pump*	16		
Apus	Apodis	Aps	*Bird of Paradise*	20		
Aquarius	Aquarii	Aqr	*Water Carrier*	13		
Aquila	Aquilae	Aql	*Eagle*	12		
Ara	Arae	Ara	*Altar*	18		
Aries	Arietis	Ari	*Ram*	8	2	
Auriga	Aurigae	Aur	*Charioteer*	3		
Boötes	Boötis	Boo	*Herdsman*	5	11	
Caelum	Caeli	Cae	*Engraving Tool*	15		
Camelopardalis	Camelopardalis	Cam	*Giraffe*	1	3	
Cancer	Cancri	Cnc	*Crab*	10	4	
Canes Venatici	Canum Venaticorum	CVn	*Hunting Dogs*	5		
Canis Major	Canis Majoris	CMa	*Greater Dog*	9	15	
Canis Minor	Canis Minoris	CMi	*Lesser Dog*	9		
Capricornus	Capricorni	Cap	*Sea Goat*	13		
Carina	Carinae	Car	*Keel*	16	20	
Cassiopeia	Cassiopeiae	Cas	*Cassiopeia*	1	2	
Centaurus	Centauri	Cen	*Centaur*	17		
Cepheus	Cephei	Cep	*Cepheus*	1		
Cetus	Ceti	Cet	*Whale*	8		
Chamaeleon	Chamaeleonis	Cha	*Chameleon*	20		
Circinus	Circini	Cir	*Pair of Compasses*	17	20	
Columba	Columbae	Col	*Dove*	15		
Coma Berenices	Coma Berenicis	Com	*Berenice's Hair*	5	11	
Corona Australis	Coronae Australis	CrA	*Southern Crown*	18		
Corona Borealis	Coronae Borealis	CrB	*Northern Crown*	5	6	
Corvus	Corvi	Crv	*Crow*	11		
Crater	Crateris	Crt	*Cup*	10		
Crux	Crucis	Cru	*Southern Cross*	16	17	20
Cygnus	Cygni	Cyg	*Swan*	7		
Delphinus	Delphini	Del	*Dolphin*	13		
Dorado	Doradus	Dor	*Goldfish*	15		
Draco	Draconis	Dra	*Dragon*	1	6	
Equuleus	Equulei	Equ	*Little Horse*	13		
Eridanus	Eridani	Eri	*River Eridanus*	8	9	14
Fornax	Fornacis	For	*Furnace*	14		
Gemini	Geminorum	Gem	*Twins*	3	9	
Grus	Gruis	Gru	*Crane*	19		
Hercules	Herculis	Her	*Hercules*	6	12	
Horologium	Horologii	Hor	*Pendulum Clock*	14		
Hydra	Hydrae	Hya	*Water Snake*	10	16	17
Hydrus	Hydri	Hyi	*Lesser Water Snake*	20		
Indus	Indi	Ind	*Indian*	19	20	

Name	Genitive	Abbreviation	Common name	Chart number(s)		
Lacerta	Lacertae	Lac	*Lizard*	7		
Leo	Leonis	Leo	*Lion*	10	4	
Leo Minor	Leonis Minoris	LMi	*Lesser Lion*	4		
Lepus	Leporis	Lep	*Hare*	9		
Libra	Librae	Lib	*Scales*	11	17	
Lupus	Lupi	Lup	*Wolf*	17	18	
Lynx	Lyncis	Lyn	*Lynx*	4		
Lyra	Lyrae	Lyr	*Lyre*	6		
Mensa	Mensae	Men	*Table Mountain*	20		
Microscopium	Microscopii	Mic	*Microscope*	19		
Monoceros	Monocerotis	Mon	*Unicorn*	9		
Musca	Muscae	Mus	*Fly*	20		
Norma	Normae	Nor	*Level*	17	18	
Octans	Octantis	Oct	*Octant*	20		
Ophiuchus	Ophiuchi	Oph	*Serpent Holder*	12	18	
Orion	Orionis	Ori	*Orion, the Hunter*	9		
Pavo	Pavonis	Pav	*Peacock*	20	18	
Pegasus	Pegasi	Peg	*Pegasus*	13	7	
Perseus	Persei	Per	*Perseus*	2	3	
Phoenix	Phoenicis	Phe	*Phoenix*	14		
Pictor	Pictoris	Pic	*Painter's Easel*	15		
Pisces	Piscium	Psc	*Fishes*	8	13	2
Piscis Austrinus	Piscis Austrini	PsA	*Southern Fish*	19		
Puppis	Puppis	Pup	*Stern*	15	9	
Pyxis	Pyxidis	Pyx	*Mariner's Compass*	16		
Reticulum	Reticuli	Ret	*Net*	14	15	
Sagitta	Sagittae	Sge	*Arrow*	6	12	
Sagittarius	Sagittarii	Sgr	*Archer*	18	12	
Scorpius	Scorpii	Sco	*Scorpion*	18	12	
Sculptor	Sculptoris	Scl	*Sculptor*	14	19	
Scutum	Scuti	Sct	*Shield*	12		
Serpens	Serpentis	Ser	*Serpent*	12	11	
Sextans	Sextantis	Sex	*Sextant*	10		
Taurus	Tauri	Tau	*Bull*	9	3	
Telescopium	Telescopii	Tel	*Telescope*	18	19	
Triangulum	Trianguli	Tri	*Triangle*	2		
Triangulum Australe	Trianguli Australis	TrA	*Southern Triangle*	20		
Tucana	Tucanae	Tuc	*Toucan*	20		
Ursa Major	Ursae Majoris	UMa	*Great Bear*	4	1	
Ursa Minor	Ursae Minoris	UMi	*Lesser Bear*	1		
Vela	Velorum	Vel	*Sail*	16		
Virgo	Virginis	Vir	*Virgin*	11		
Volans	Volantis	Vol	*Flying Fish*	20		
Vulpecula	Vulpeculae	Vul	*Fox*	6	7	

Variable stars

The brightness of many stars varies over longer or shorter periods of time. The most common reason for this is that the size of the star actually changes: the star pulsates. A well-known type of pulsating star is the Cepheid, named after Delta (δ) Cephei, a yellow supergiant, which regularly pulsates every few days or weeks. The Cepheids are divided into two classes: the classical (Cep) and the Population II (Cep W) Cepheids. They are important to astronomers because there is a relation between their period and luminosity. The brighter a Cepheid, the longer the period. When the period is measured, we know the real luminosity (absolute brightness) of the star. By comparing this with the amount of light we actually receive (apparent brightness) we have an important tool for calculating distance.

Another type of pulsating variable is named after the prototype Omicron (o) Ceti or Mira (M), a red giant. These variables do not have a strict period. Several other types of pulsating variables are also named after prototypes, like U Gem, R CrB, or RR Lyr. There is also a completely different type of variable star: the eclipsing variable. An eclipsing variable star is a double star in mutual orbit and one component periodically moves in front, or behind the other, causing a drop in the total amount of light we receive. The best-known example of this type is Beta (β) Persei or Algol. Eclipsing variables are referred to in the tables as type E, subdivided by the shape of their light curves into E, EA, EB and EW.

A further group is termed the eruptive variables. These undergo a very sudden and very large increase in brightness. The best known of these are the novae and the supernovae. A nova is a very close double star, in which one component is a white dwarf: a small, but very compact star. Gas from the other component flows into the white dwarf and it ignites in a huge explosion. The brightness of the star increases temporarily by thousands of times. Some novae erupt more than once. These are known as recurrent novae.

A supernova is even more spectacular. It is the catastrophic death of a very hot star. The star's life ends when it blows itself up and for a short period it shines millions of times brighter than it was. After

the star has faded again the outer shells of the star form a slowly expanding nebula. The Crab nebula (M1) in Taurus (the Bull) is one example. On the atlas all variable stars with a maximum of magnitude 6.5 or brighter are plotted.

Double stars

The majority of stars are double or multiple stars. Sometimes two stars appear very close in the sky, but are only on the same line of sight, while their distances differ considerably. These are called optical double stars. Real physical double stars belong together and are also called binaries. They are tied together by gravity and are in mutual orbit. The same goes for triple, quadruple and even larger families of stars. Their apparent separation is measured in minutes and seconds of arc. One degree (°) equals 60 minutes (') and one minute again equals 60 seconds ("). The separation in the tables (column with heading Sep) is given in seconds of arc. There is also a column with the heading PA, meaning position angle. This gives the angular position of the fainter component in relation to the brighter one. The angle is measured from the north, eastward. Keep in mind that in the sky east and west are reversed. So, when north is up, east is to the left! Consequently, the position angle is measured anti-clockwise.

All double stars with an integrated (combined) magnitude of 6.5 or brighter are plotted on the charts. The tables contain only a selection of the finest targets.

Open and globular clusters

Open and globular star clusters appear on the charts as yellow disks. Open clusters are shown with a dotted outline and globular clusters are shown as solid circles with a cross through the center. Open clusters are usually found near the plane of the Milky Way, and so are in or close to the soft blue areas on the charts, representing the brightest parts of the Milky Way. Open clusters are groups of young stars, often hot and bluish, and their individual stars can easily be seen in a small telescope or sometimes with the naked eye. The last column of the table gives the approximate number of stars in the cluster.

Globular clusters are quite different. They contain larger numbers of stars and are much more compact. They are found outside the galaxy plane and the stars are older. All clusters down to magnitude 10 are plotted on the charts. The diameter (Diam) in the table is given in minutes (') of arc.

In the first column you will find the designation of the cluster. First the NGC numbers are listed (NGC = New General Catalogue) and on the charts these numbers are found without the prefix NGC. Below these come the IC numbers (IC = Index Catalogue). These have the prefix I. on the charts. Clusters from other catalogues follow on at the end of the list. Alternative names are in the last column. The same goes for nebulae and galaxies.

Planetary and diffuse nebulae

Planetary nebulae have nothing to do with planets. The name arose from their disk-like appearance. They are almost spherical cast-off shells of gas from very hot stars, late in their life-spans. Often the ionized gas has a greenish color. Examples are the Ring nebula (M57) in Lyra (the Lyre) and the Helix nebula (NGC 7293) in Aquarius (the Water Carrier).

Diffuse nebulae are areas of the raw materials, dust and gas, from which stars are born. These diffuse nebulae are also found along the spiral arms of the Milky Way, and are visible in other nearby galaxies. Both planetary and diffuse nebulae are printed in soft green on the atlas charts.

Galaxies

The red ovals on the charts are the most remote objects: the galaxies. Galaxies are huge systems of stars, clusters and nebulae, like our own Milky Way. There are several types of galaxies: the elliptical (E), spiral (S), barred spiral (SB) and irregular. The E type is subdivided according to shape. E0 for the almost spherical, to E7 for the flattened lens shape. The S and SB types are subdivided according to how tight the spiral arms are wound (Figure 4).

The Galactic Equator

On the charts the Galactic Equator is drawn as a line of dots and dashes. It represents the projection of the plane of our galaxy on the stellar sphere. Every ten degrees of galactic longitude is marked along this equator. The center of our Milky Way is at 0°.

The Ecliptic

The Ecliptic is the projection of the Earth's orbit around the Sun, or the yearly path of the Sun along the sky, caused by the orbital movement of the Earth. It is shown on the charts as a broken line. The Moon and the planets are always close to the Ecliptic.

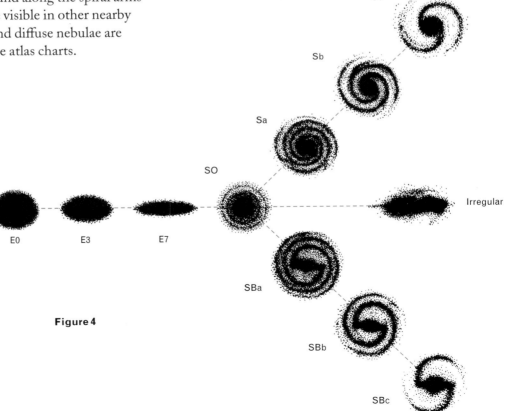

Figure 4

Abbreviations used in the star chart tables

Lists of telescopic objects are given for each map. It is clearly impossible to give a complete list, and the objects selected are those which are shown on the maps and which are within the range of telescopes of the size usually owned by amateurs. All positions are given for epoch 2000.0.

Variable stars

Range, type, period and spectrum are given.

Cep	classical Cepheid
CW	type II Cepheid
E	eclipsing binary
EA	Algol type
EB	Beta Lyrae type
EW	W Ursae Majoris type
M	Mira (long-period) type
SR	semi-regular
Irr	irregular
R CrB	R Coronae Borealis type
δ Sct	delta Scuti type
ZA	Z Andromedae type
RN	recurrent nova
N	nova
RV Tau	RV Tauri type

Some stars indicated as variable in the maps have visual ranges of 0.3 mag or less. These are not included in the lists.

Double stars

PA	position angle, from north (0°) through east (90°), south (180°) and west (270°). With binary stars, the values given are those for approximately the year 1990, though with binaries of reasonably short period these data will alter quite quickly
Sep	separation, in seconds of arc. The data are for the most recent available measurements

Open clusters

NGC	New General Catalogue
Diam	diameter, in minutes of arc
Mag	approximate total visual magnitude
N*	approximate number of stars

Globular clusters

NGC	New General Catalogue
Diam	diameter, in minutes of arc
Mag	approximate integrated visual magnitude

Planetary nebulae

NGC	New General Catalogue
Diam	diameter, in seconds of arc
Mag	integrated photographic magnitude
Mag*	photoelectric magnitude of the central star

Nebulae

NGC	New General Catalogue
Diam	approximate maximum and minimum angular dimensions, in minutes of arc
Mag*	approximate magnitude of the illuminating star

Galaxies

NGC	New General Catalogue
Mag	approximate total integrated visual magnitude
Diam	major and minor diameters, in minutes of arc (again, bound to be somewhat arbitrary)
Type	the type according to the classical Hubble system

The lists apply to the main regions of each map. Objects such as old novae are not included, since in general they lie beyond the range of any but very powerful telescopes; also excluded are double stars of very small separation (below 0.1) or with very faint secondary components.

The constellation abbreviations follow the official IAU practice. (See the list of constellations in Table C.)

Index to the atlas charts

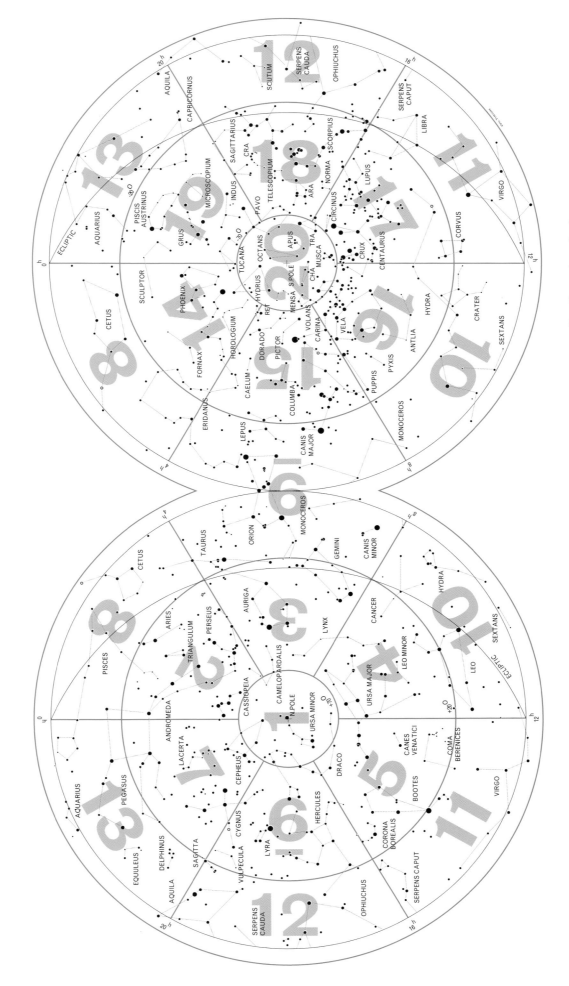

Southern hemisphere

Northern hemisphere

Chart 1 *Far north; declination above +70°*

Variable stars

		RA h	m	Dec °	Range	Type	Period d	Spectrum
VZ	Cam	07	31.1	+82 25	4.8–5.2	**SR**	23.7	M
YZ	Cas	00	45.7	+74 59	5.7–6.1	**EA**	4.47	A+F
UX	Dra	19	21.6	+76 34	5.9–7.1	**SR**	168	N

Double stars

		RA h	m	Dec	PA	Sep	Mag
48	**Cas**	02	02.0	+70 54	234	0.9	4.7, 6.9 Binary, 60 y
49	**Cas**	02	05.5	+76 07	246	5.4	5.3,12.3
β	**Cep**	21	28.7	+70 34	249	13.3	3.2, 7.9
κ	**Cep**	20	08.9	+77 43	122	7.4	4.4, 8.4
π	**Cep**	23	07.9	+75 23	346	1.2	4.6, 6.6 Slow binary
ε	**Dra**	19	48.2	+70 16	015	3.1	3.8, 7.4
ψ	**Dra**	17	41.9	+72 09	015	30.3	4.9, 6.1
α	**UMi**	02	31.8	+89 16	218	18.4	2.0, 9.0 Polaris
5	**UMi**	14	27.5	+75 42	AB 124	21.7	4.3,13.3
					AC 131	58.8	9.8

Open clusters

NGC		RA		Dec	Diam	Mag	N*
188	**Cep**	00	44.4	+85 20	14	8.1	120

Planetary nebulae

NGC		RA		Dec	Diam	Mag	Mag*
IC 3568	**Cam**	12	32.9	+82 33	6	11.6	12.3
40	**Cep**	00	13.0	+72 32	37	10.7	11.6

Galaxies

NGC		RA		Dec	Mag	Diam	Type
2146	**Cam**	06	18.7	+78 21	10.5	6.0×3.8	**SBb**
2655	**Cam**	08	55.6	+78 13	10.1	5.1×4.4	**SBa**
2715	**Cam**	09	08.1	+78 05	11.4	5.0×1.9	**Sc**
2985	**UMa**	09	50.4	+72 17	10.5	4.5×3.4	**Sb**
3147	**Dra**	10	16.9	+73 24	10.6	4.0×3.5	**Sb**
6503	**Dra**	17	49.4	+70 09	10.2	6.2×2.3	**Sb**

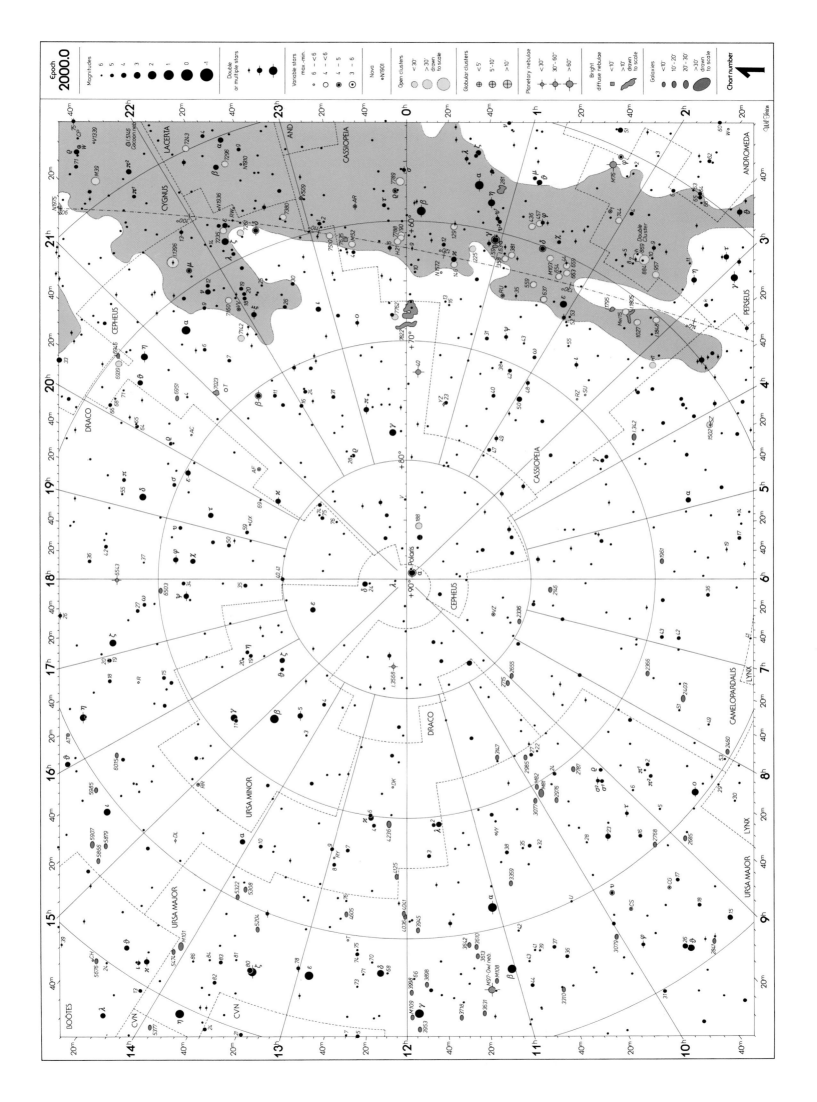

Chart 2 RA 0h to 4h. Dec +20° to 70°

Variable stars

		RA h m	Dec °	Range	Type	Period d	Spectrum
R	And	00 24.0	+38 35	5.8–14.9	M	409.3	M
W	And	02 17.6	+44 18	6.7–14.6	M	395.9	S
α	Cas	00 40.5	+56 32	?2.2– 2.5	Suspected	—	K
γ	Cas	00 56.7	+60 43	1.6– 3.3	Irr	—	B
RZ	Cas	02 48.9	+69 38	6.2– 7.7	EA	1.19	A
β	Per	03 08.2	+40 57	2.2– 3.4	EA	2.87	B+G Algol
σ	Per	03 05.2	+38 50	3.0– 4.0	SR	50	M
X	Per	03 55.4	+31 03	6.0– 7.0	Irr	—	09.05
R	Tri	02 37.0	+34 16	5.4–12.6	M	266.5	M

Double stars

		RA h m	Dec °	PA	Sep	Mag
γ	And	02 03.9	+42 20	063	9.8	2.3, 4.8
γ²	And			106	0.5	5.5, 6.3 Binary, 61.1 y
δ	And	00 39.3	+30 52	298	28.7	3.3,12.4
π	And	00 36.9	+33 43	173	35.9	4.4, 8.6
φ	And	01 09.5	+47 15	130	0.5	4.6, 5.5 Binary, 372 y
ϖ	And	01 27.7	+45 24	AB 117 / AC 111	2.0 / 119.0	4.8,11.5 / 10.2
55	And	01 52.2	+40 44	356	59.8	5.6,10.9
ε	Ari	02 59.2	+21 20	191	1.5	5.2, 5.5 C:12.7,146"
30	Ari	02 37.0	+24 39	274	38.6	6.6, 7.4
33	Ari	02 40.7	+27 04	000	28.6	5.5, 8.4
41	Ari	02 50.0	+27 38	AB 277 / AC 213	24.6 / 31.3	3.6,10.7 / 10.5
α	Cas	00 40.5	+56 32	275	19.8	2.3v?,13.7 Optical
γ	Cas	00 56.7	+60 43	248	2.1	2v, 11.2
η	Cas	00 49.1	+57 49	293	12.2	3.4, 7.5 Binary, 480 y
τ	Cas	02 29.1	+67 24	232	2.4	4.6, 6.9 Binary, 840 y
λ	Cas	00 31.8	+54 31	176	0.5	5.3, 5.6
ψ	Cas	01 25.9	+68 08	AC 113 / AD 118	25.0 / 22.8	4.7, 9.6 / 9.7
35	Cas	01 21.1	+64 40	344	55.5	6.3, 8.7
γ	Per	03 04.8	+53 30	326	57.0	2.9,10.6
ε	Per	03 57.9	+40 01	010	8.8	2.9, 8.1
ζ	Per	03 54.1	+31 53	AB 208 / AC 286 / AD 195 / AE 185	12.9 / 32.8 / 94.2 / 120.3	2.9, 9.5 / 11.3 / 9.5 / 10.2
η	Per	02 50.7	+55 54	300	28.3	3.3, 8.5
θ	Per	02 44.2	+49 14	215	19.8	4.1, 9.9 Binary, 2720 y
o	Per	03 44.3	+32 17	037	1.0	3.8, 8.3
τ	Per	02 54.3	+52 46	106	51.7	3.0,10.6
ψ	Psc	01 05.6	+21 28	159	30.0	5.6, 5.8
6	Tri	02 12.4	+30 18	071	3.9	5.3, 6.9

Open clusters

NGC		RA	Dec	Diam	Mag	N*	
129	Cas	00 29.9	+60 14	21	6.5	35	Contains DL Cas
133	Cas	00 31.2	+63 22	7	9.4	5	
146	Cas	00 33.1	+63 18	7	9.1	20	
381	Cas	01 08.3	+61 35	6	9.3	50	
436	Cas	01 15.6	+58 49	6	8.8	30	
457	Cas	01 19.1	+58 20	13	6.4	80	φ Cas cluster
559	Cas	01 29.5	+63 18	4.4	9.5	60	
581	Cas	01 33.2	+60 42	6	7.4	25	M103
637	Cas	01 42.9	+64 00	3.5	8.2	20	
654	Cas	01 44.1	+61 53	5	6.5	60	
659	Cas	01 44.2	+60 42	5	7.9	40	
663	Cas	01 46.0	+61 15	16	7.1	80	
744	Per	01 58.4	+55 29	11	7.9	20	
752	And	01 57.8	+37 41	50	5.7	60	
869	Per	02 19.0	+57 09	30	4.3	200	} Sword-handle of Perseus
884	Per	02 22.4	+57 07	30	4.4	150	}
957	Per	02 33.6	+57 32	11	7.6	30	
1027	Cas	02 42.7	+61 33	20	6.7	40	
1039	Per	02 42.0	+42 47	35	5.2	60	M34
1245	Per	03 14.7	+47 15	10	8.4	200	
1432/5	Tau	03 47.0	+24 07	110	1.2	300+	M45 Pleiades
1444	Per	03 49.4	+52 40	4	6.6	—	
IC 1805	Cas	02 32.7	+61 27	22	6.5	40	

Planetary nebulae

NGC		RA	Dec	Diam	Mag	Mag*	
650–1	Per	01 42.4	+51 34	65×290	12.2	17	Little Dumbbell M76

Nebulae

NGC		RA	Dec	Diam	Mag*
281	Cas	00 52.8	+56 36	35×30	8
IC 1805	Cas	02 33.4	+61 26	60×60	—
IC 1848	Cas	02 51.3	+60 25	60×30	—

Galaxies

NGC		RA	Dec	Mag	Diam	Type	
147	Cas	00 33.2	+48 30	9.3	12.9× 8.1	dE4	
185	Cas	00 39.0	+48 20	9.2	11.5× 9.8	dE0	
205	And	00 40.4	+41 41	8.0	17.4× 9.8	E6	M110 Companion to M31
221	And	00 42.7	+40 52	8.2	7.6× 5.8	E2	M32 Companion to M31
224	And	00 42.7	+41 16	3.5	178 ×63	Sb	M31
598	Tri	01 33.9	+30 39	5.7	62 ×39	Sc	M33 Pinwheel
891	And	02 22.6	+42 21	9.9	13.5× 2.8	Sb	
925	Tri	02 27.3	+33 35	10.0	9.8× 6.0	SBc	
976	Ari	02 34.0	+20 59	12.4	1.7× 1.5	Sb	
1003	Per	02 39.3	+40 52	11.5	5.4× 2.1	Sc	
1023	Per	02 40.0	+39 04	9.5	8.7× 3.3	E7	
IC 342	Cam	03 46.8	+68 06	9.2	17.8×17.4	SBc	

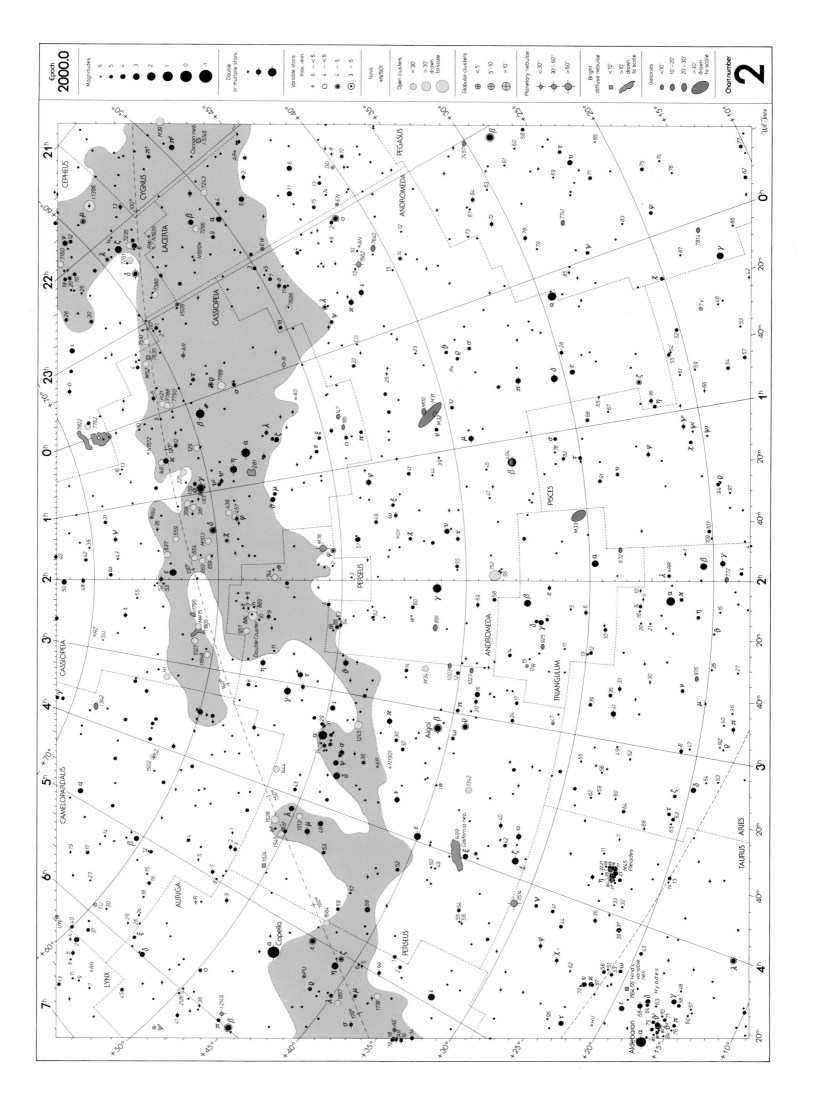

Chart 3 RA 4h to 8h. Dec +20° to 70°

Variable stars

		RA h m	Dec °	Range	Type	Period d	Spectrum
ε	Aur	05 02.0	+43 49	2.9– 3.8	E	9892	A–F
ζ	Aur	05 02.5	+41 05	3.7– 4.0	E	972.1	K+B
R	Aur	05 17.3	+53 35	6.7–13.9	M	457.5	M
RT	Aur	06 28.6	+30 30	5.0– 5.8	Cep	3.73	F–G
UU	Aur	06 36.5	+38 27	5.1– 6.8	SR	234	N
WW	Aur	06 32.5	+32 27	5.8– 6.5	EA	2.53	A+A
ζ	Gem	07 04.1	+20 34	3.7– 4.1	Cep	10.15	F–G
η	Gem	06 14.9	+22 30	3.2– 3.9	SR	233	M
R	Gem	07 07.4	+22 42	6.0–14.0	M	369.8	S
BU	Gem	06 12.3	+22 54	5.7– 7.5	Irr	—	M
RR	Lyn	06 26.4	+56 17	5.6– 6.0	EA	9.95	A
U	Ori	05 55.8	+20 10	4.8–12.6	M	372.4	M
HU	Tau	04 38.3	+20 41	5.9– 6.7	EA	2.06	A

Double stars

		RA	Dec	Pair	PA	Sep	Mag
δ	Aur	05 59.5	+54 17	AB 271		115.4	3.7, 9.5
				AC 067		197.1	9.5
θ	Aur	05 59.7	+37 13	AB 313		3.6	2.6, 7.1
				AC 297		50.0	10.6
ν	Aur	05 51.5	+39 09	206		54.6	4.9, 9.3
ϖ	Aur	04 59.3	+37 53	359		5.4	5.0, 8.0
R	Aur	05 17.3	+53 35	339		47.5	var, 8.6
14	Aur	05 15.4	+32 41	AB 352		11.1	5.1,11.1
				AC 226		14.6	5.1, 7.4
26	Aur	05 38.6	+30 30	AB 031		0.1	6.0, 6.3 Binary, 53.2 y
				AB+C 267		12.4	6.0, 8.0
1	Cam	04 32.0	+53 55	308		10.3	5.7, 6.8
α	Gem	07 34.6	+31 53	AB 088		2.5	1.9, 2.9 Binary, 420 y
				AC 164		72.5	1.6, 8.8
δ	Gem	07 20.1	+21 59	223		6.0	3.5, 8.2 Binary, 1200 y
ε	Gem	06 43.9	+25 08	094		110.3	3.0, 9.0
η	Gem	06 14.9	+22 30	266		1.4	var, 8.8 Binary, 474 y
κ	Gem	07 44.4	+24 24	240		7.1	3.6, 8.1
μ	Gem	06 22.9	+22 31	077		72.7	3.2, 9.8
ν	Gem	06 29.0	+20 13	329		112.5	4.2, 8.7
π	Gem	07 47.5	+33 25	AB 214		21.0	5.1,11.2
				AC 341		91.9	10.2
ρ	Gem	07 29.1	+31 47	AB 008		3.4	4.2,12.5
				AC 291		213.7	10.6
				CD 267		104.1	12.2
3	Gem	06 09.7	+23 07	339		0.5	5.8, 9.9
65	Gem	07 29.8	+27 55	288		12.8	5.0,13.5
70	Gem	07 38.5	+35 03	AB 191		100.4	5.6,11.6
				AC 100		162.0	10.6
				CD 242		1.6	11.6
				CE 207		17.5	14.1
4	Lyn	06 22.1	+59 22	AB 124		0.8	6.2, 7.7
				AB+C 096		26.2	12.9
				AB+D 356		100.4	11.0
12	Lyn	06 46.2	+59 27	AB 070		1.7	5.4, 6.0 Binary, 699 y
				AC 308		8.7	7.3
				AB 256		170.0	10.6
14	Lyn	06 53.1	+59 27	AB 260		0.4	5.7, 6.9 Binary, 480 y
				AB+C 122		181.0	11.0
19	Lyn	07 22.9	+55 17	AB 315		14.8	5.6, 6.5
				AD 003		214.9	8.9
				BC 287		74.2	
24	Lyn	07 43.0	+58 43	320		54.7	5.0, 9.5
μ	Per	04 14.9	+48 25	349		14.8	4.1,11.6
56	Per	04 24.6	+33 58	022		4.2	5.9, 8.7
κ+67	Tau	04 25.4	+22 18	173		339.0	4.2, 5.3
φ	Tau	04 20.4	+27 21	250		52.1	5.0, 8.4
χ	Tau	04 22.6	+25 38	024		19.4	5.5, 7.6
103	Tau	05 08.1	+24 16	AB 150		13.3	5.5,12.0
				AC 197		35.3	8.6
118	Tau	05 29.3	+25 09	AB 204		4.8	5.8, 6.6
				AC 099		141.3	11.6

Open clusters

NGC		RA	Dec	Diam	Mag	N*	
1502	Cam	04 07.7	+62 20	8	5.7	45	
1513	Per	04 10.0	+49 31	9	8.4	50	
1528	Per	04 15.4	+51 14	40	6.4	40	
1545	Per	04 20.9	+50 15	8	6.2	20	
1664	Aur	04 51.1	+43 42	18	7.6	—	
1746	Tau	05 03.6	+23 49	42	6.1	20	
1778	Aur	05 08.1	+37 03	7	7.7	25	
1857	Aur	05 20.2	+39 21	6	7.0	40	
1893	Aur	05 22.7	+33 24	11	7.5	60	
1912	Aur	05 28.7	+35 50	21	6.4	100+	M38
1960	Aur	05 36.1	+34 08	12	6.0	60	M36
2099	Aur	05 52.4	+32 33	24	5.6	150	M37
2126	Aur	06 03.0	+49 54	6	10.2	40	
2129	Gem	06 01.0	+23 18	7	6.7	40	
2168	Gem	06 08.9	+24 20	28	5.1	200	M35
2175	Ori	06 09.8	+20 19	18	6.8	60	
2266	Gem	06 43.2	+26 58	7	9.5	50	
2281	Aur	06 49.3	+41 04	15	5.4	30	
2420	Gem	07 38.5	+21 34	10	8.3	100	
IC 2157	Gem	06 05.0	+24 00	7	8.4	20	

Planetary nebulae

NGC		RA	Dec	Diam	Mag	Mag*	
1514	Tau	04 09.2	+30 47	114	10	9.4	
2392	Gem	07 29.2	+20 55	13×44	10	10.5	Eskimo Nebula

Nebulae

NGC		RA	Dec	Diam	Mag	
1499	Per	04 00.7	+36 37	145×40	4	California Nebula
1952	Tau	05 34.5	+22 01	6×4	16	M1 Crab Nebula: SNR
IC 405	Aur	05 16.2	+34 16	30×19	6v	AE Aur: Flaming Star Nebula

Galaxies

NGC		RA	Dec	Mag	Diam	Type
1961	Cam	05 42.1	+69 23	11.1	4.3× 3.0	Sbp
2366	Cam	07 28.9	+69 13	10.9	7.6× 3.5	Irr
2403	Cam	07 36.9	+65 36	8.4	17.8×11.0	Sc
2460	Cam	07 56.9	+60 21	11.7	2.9× 2.2	Sb

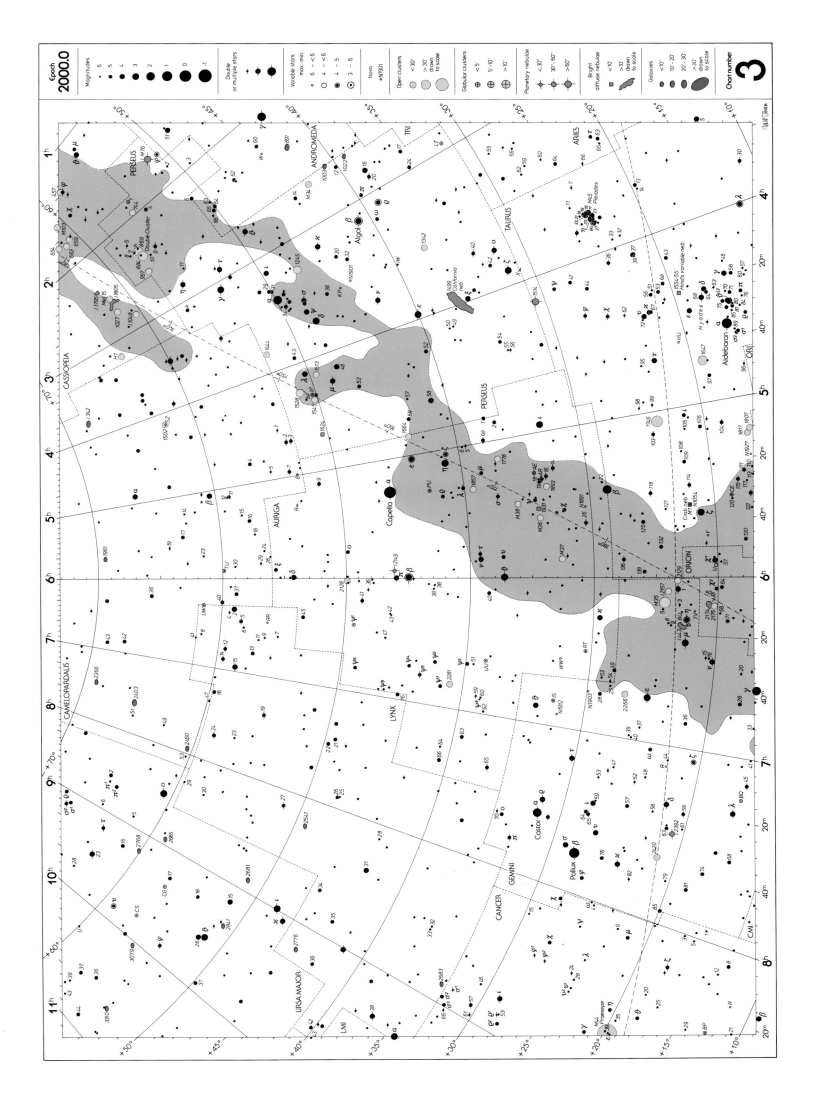

Chart 4 RA 8h to 12h. Dec +20° to +70°

Variable stars

		RA h m	Dec °	Range	Type	Period d	Spectrum
RS	Cnc	09 10.6	+30 58	6.2– 7.7	SR	120	M
UY	Leo	10 29.4	+23 04	9.5–11.0	Irr	—	M
R	LMi	09 45.6	+34 31	6.3–13.2	M	371.9	M
ST	UMa	11 27.8	+45 11	7.7– 9.5	SR	81	M
VY	UMa	10 45.7	+67 25	5.9– 6.5	Irr	—	N

Double stars

		RA	Dec	PA	Sep	Mag	
β	LMi	10 27.9	+36 42	250	0.2	4.6, 6.1	Binary, 37.2 y
11	LMi	09 35.7	+35 49	031	2.0	5.4,13.9	
40	LMi	10 43.0	+26 20	112	18.4	5.5,12.5	
38	Lyn	09 18.8	+36 48	AB 229	2.7	3.9, 6.6	
				BC 212	87.7	10.8	
				BD 256	177.9	10.7	
α	UMa	11 03.7	+61 45	283	0.7	1.9, 4.9	Binary, 44.7 y
ι	UMa	08 59.2	+48 02	100	1.8	3.1,10.2	Binary, 818 y
κ	UMa	09 03.6	+47 09	258	0.1	4.2, 4.4	Binary, 70y
o	UMa	08 30.3	+60 43	192	7.1	3.4,15	
ν	UMa	11 18.5	+33 06	147	7.2	3.5, 9.9	
ξ	UMa	11 18.2	+31 32	060	1.3	4.3, 4.8	Binary, 59.8 y
φ	UMa	09 52.1	+54 04	188	0.2	5.3, 5.4	Binary, 105.5 y
23	UMa	09 31.5	+63 04	AB 270	22.7	3.7, 8.9	
				AC 231	99.6	10.4	
57	UMa	11 29.1	+39 20	AB 359	5.4	5.3, 8.3	
				AC 009	216.5	11.5	

Planetary nebula

NGC	RA	Dec	Diam	Mag	Mag*	
3587	11 14.8	+55 01	194	12.0	15.9	M97 Owl Nebula

Galaxies

NGC		RA	Dec	Mag	Diam	Type	
2541	Lyn	08 14.7	+49 04	11.7	6.6× 3.5	S	
2681	UMa	08 53.5	+51 19	10.3	3.8× 3.5	Sa	
2683	Lyn	08 52.7	+33 25	9.7	9.3× 2.5	Sb	
2685	UMa	08 55.6	+58 44	11.0	5.2× 3.0	Sbp	
2768	UMa	09 11.6	+60 02	10.0	6.3× 2.8	E5	
2776	Lyn	09 12.2	+44 57	11.6	2.9× 2.7	Sc	
2787	UMa	09 19.3	+69 12	10.8	3.4× 2.3	Sap	
2841	UMa	09 22.0	+50 58	9.3	8.1× 3.8	Sb	
2976	UMa	09 47.3	+67 55	10.1	4.9× 2.5	Scp	
3003	LMi	09 48.6	+33 25	11.7	5.9× 1.7	SBc	
3031	UMa	09 55.6	+69 04	6.9	25.7×14.1	SB	M81
3034	UMa	09 55.8	+69 41	8.4	11.2× 4.6	Pec	M82
3077	UMa	10 03.3	+68 44	9.8	4.6× 3.6	E2p	
3079	UMa	10 02.0	+55 41	10.6	7.6× 1.7	Sb	
3184	UMa	10 18.3	+41 25	9.7	6.9× 6.8	Sc	
3190	Leo	10 18.1	+21 50	11.0	4.6× 1.8	Sb	
3198	UMa	10 19.9	+45 33	10.4	8.3× 3.7	Sc	
3245	LMi	10 27.3	+28 30	10.8	3.2× 1.9	E5	
3254	LMi	10 29.3	+29 30	11.5	5.1× 1.9	Sb	
3294	LMi	10 36.3	+37 20	11.7	3.3× 1.8	Sc	
3310	UMa	10 38.7	+53 30	10.9	3.6× 3.0	SBc	
3344	LMi	10 43.5	+24 55	9.9	6.9× 6.5	Sc	
3359	UMa	10 46.6	+63 13	10.4	6.8× 4.3	SBc	
3414	LMi	10 51.3	+27 59	10.7	3.6× 2.7	SBa	
3430	LMi	10 52.2	+32 57	11.5	3.9× 2.3	Sc	
3432	LMi	10 52.5	+36 37	11.2	6.2× 1.5	SB	
3486	LMi	11 00.4	+28 58	10.3	6.9× 5.4	Sc	
3610	UMa	11 18.4	+58 47	10.7	3.2× 2.5	E2p	
3631	UMa	11 21.0	+53 10	10.4	4.6× 4.1	Sc	
3646	Leo	11 21.7	+20 10	11.2	3.9× 2.6	Sc	
3675	UMa	11 26.1	+43 35	10.9	5.9× 3.2	Sb	
3687	UMa	11 28.0	+29 31	12.6	2.0× 2.0	Sb	
3718	UMa	11 32.6	+53 04	10.5	8.7× 4.5	SBap	
3726	UMa	11 33.3	+47 02	10.4	6.0× 4.5	Sc	
3877	UMa	11 46.1	+47 30	11.6	5.4× 1.5	Sb	
3898	UMa	11 49.2	+56 05	10.8	4.4× 2.6	Sb	
3945	UMa	11 53.2	+60 41	10.6	5.5× 3.6	SBa	
3949	UMa	11 53.7	+47 52	11.0	3.0× 1.8	Sb	
3953	UMa	11 53.8	+55 20	10.1	6.6× 3.6	Sb	
3998	UMa	11 57.9	+55 27	10.6	3.1× 2.5	E2p	
4026	UMa	11 59.4	+50 58	11.7	5.1× 1.4	S0	

Chart number 4

Chart 5 · RA 12h to 16h. Dec +20° to +70°

Variable stars

		RA h m	Dec °	Range	Type	Period d	Spectrum
W	Boö	14 43.4	+26 32	4.7– 5.4	SR	450	M
i(44)	Boö	15 03.8	+47 39	6.5– 7.1	EW	0.27	G+G
R	CVn	13 49.0	+39 33	6.5–12.9	M	328.5	M
Y	CVn	12 45.1	+45 26	7.4–10.0	SR	157	N La Superba
TU	CVn	12 54.9	+47 12	5.6– 6.6	SR	50	M
FS	Com	13 06.4	+22 37	5.3– 6.1	SR	58	M (FZ in map)
R	CrB	15 48.6	+28 09	5.7–15	RCrB	—	F8p
T	CrB	15 59.5	+25 55	2.0–10.8	RN	29 000?	M+P(Q) Blaze Star
RY	Dra	12 56.4	+66 00	5.6– 8.0	SR	173	N
T	UMa	12 36.4	+59 29	6.6–13.4	M	265.5	M
RR	UMi	14 57.6	+65 56	6.1– 6.5	SR?	40	M

Double stars

		RA	Dec	PA	Sep	Mag	
γ	Boö	14 32.1	+38 19	111	33.4	3.0,12.7	
ε	Boö	14 45.0	+27 04	339	2.8	2.5, 4.9	
κ	Boö	13 13.5	+51 47	236	13.4	4.6, 6.6	
μ	Boö	15 24.5	+37 23	171	108.3	4.3, 7.0	
39	Boö	14 49.7	+48 43	045	2.9	6.2, 6.9	
i(44)	Boö	15 03.8	+47 39	040	1.0	5.3v,6.2 Binary, 225 y	
α	CVn	12 56.0	+38 19	229	19.4	2.9, 5.5 Cor Caroli	
2	CVn	12 16.1	+40 40	260	11.4	5.8, 8.1	
ε	CrB	15 57.6	+26 53	AB 003 / AC 174	1.8 / 101.4	4.2,12.6 / 11.5	
η	CrB	15 23.2	+30 17	AB 030 / AC 012 / AB+D 047	1.0 / 57.7 / 215.0	5.6, 5.9 Binary, 41.6 y / 12.5 / 10.9	
ξ	UMa	13 23.9	+54 56	AB 152 / AC 071	14.4 / 708.7	2.3, 4.0 Mizar / 2.1, 4.0 Mizar/Alcor	
78	UMa	13 00.7	+56 22	057	1.5	5.0, 7.4 Binary, 116 y	

Globular cluster

NGC		RA	Dec	Diam	Mag	
5272	CVn	13 42.2	+28 23	16.2	6.4	M3

Galaxies

NGC		RA	Dec	Mag	Diam	Type	
4036	UMa	12 01.4	+62 54	10.6	4.5 × 2.0	E6	
4041	UMa	12 02.2	+62 08	11.1	2.8 × 2.7	Sc	
4051	UMa	12 03.2	+44 32	10.3	5.0 × 4.0	Sc	
4062	UMa	12 04.1	+31 54	11.2	4.3 × 2.0	Sb	
4088	UMa	12 05.6	+50 33	10.5	5.8 × 2.5	Sc	
4096	UMa	12 06.0	+47 29	10.6	6.5 × 2.0	Sc	
4100	UMa	12 06.2	+49 35	11.5	5.2 × 1.9	Sb	
4111	CVn	12 07.1	+43 04	10.8	4.8 × 1.1	SO	
4125	Dra	12 08.1	+65 11	9.8	5.1 × 3.2	E5p	
4136	Com	12 09.3	+29 56	11.7	4.1 × 3.9	Sc	
4138	CVn	12 09.5	+43 41	12.3	2.9 × 1.9	E4	
4145	CVn	12 10.0	+39 53	11.0	5.8 × 4.4	Sc	
4151	CVn	12 10.5	+39 24	10.4	5.9 × 4.4	Pec	
4214	CVn	12 15.6	+36 20	9.8	7.9 × 6.3	Irr	
4217	CVn	12 15.8	+47 06	11.9	5.5 × 1.8	Sb	
4236	Dra	12 16.7	+69 28	9.7	18.6 × 6.9	Sb	
4242	CVn	12 17.5	+45 37	11.0	4.8 × 3.8	S	
4244	CVn	12 17.5	+37 49	10.2	16.2 × 2.5	S	
4251	Com	12 18.1	+28 10	11.6	4.2 × 1.9	E7	
4258	CVn	12 19.0	+47 18	8.3	18.2 × 7.9	Sb	M106
4278	Com	12 20.1	+29 17	10.2	3.6 × 3.5	E1	
4314	Com	12 22.6	+29 53	10.5	4.8 × 4.3	SBa	
4395	CVn	12 25.8	+33 33	10.1	12.9 × 11.0	S	
4448	Com	12 28.2	+28 37	11.1	4.0 × 1.6	Sb	
4449	CVn	12 28.1	+44 06	9.4	5.1 × 3.7	Irr	
4490	CVn	12 30.6	+41 38	9.8	5.9 × 3.1	Sc	
4494	Com	12 31.4	+25 47	9.9	4.8 × 3.8	Sb	
4559	Com	12 36.0	+27 58	9.8	10.5 × 4.9	Sc	
4565	Com	12 36.3	+25 59	9.6	16.2 × 2.8	Sb	
4605	UMa	12 40.0	+61 37	11.0	5.5 × 2.3	SBep	
4618	CVn	12 41.5	+41 09	10.8	4.4 × 3.8	Sc	
4631	CVn	12 42.1	+32 32	9.3	15.1 × 3.3	Sc	
4656–7	CVn	12 44.0	+32 10	10.4	13.8 × 3.3	Sc	
4725	Com	12 50.4	+25 30	9.2	11.0 × 7.9	SBp	
4736	CVn	12 50.9	+41 07	8.2	11.0 × 9.1	Sb	M94
4826	Com	12 56.7	+21 41	8.5	9.3 × 5.4	Sab	M64 Black-Eye
5005	CVn	13 10.9	+37 03	9.8	5.4 × 2.7	Sb	
5033	CVn	13 13.4	+36 36	10.1	10.5 × 5.6	Sb	
5055	CVn	13 15.8	+42 02	8.6	12.3 × 7.6	Sb	M63
5112	CVn	13 21.9	+38 44	11.9	3.9 × 2.9	Sc	
5194	CVn	13 29.9	+47 12	8.4	11.0 × 7.8	Sc	M51 Whirlpool
5195	CVn	13 30.0	+47 16	9.3	5.4 × 4.3	Pec	Companion of M51
5308	UMa	13 47.0	+60 58	11.3	3.5 × 0.8	SO	
5322	UMa	13 49.3	+60 12	10.0	5.5 × 3.9	E2	
5371	CVn	13 55.7	+40 28	10.7	4.4 × 3.6	Sb	
5457	UMa	14 03.2	+54 21	7.7	26.9 × 26.3	Sc	M101 Pinwheel
5475	UMa	14 05.2	+55 45	13.4 photo.	2.2 × 0.6	Sa	
5676	Boö	14 32.8	+49 28	10.9	3.9 × 2.0	Sc	
5866	Dra	15 06.5	+55 46	10.0	5.2 × 2.3	E6p	
5879	Dra	15 09.8	+57 00	11.5	4.4 × 1.7	Sb	
5907	Dra	15 15.9	+56 19	10.4	12.3 × 1.8	Sb	
5985	Dra	15 39.6	+59 20	11.0	5.5 × 3.2	Sb	
6015	Dra	15 51.4	+62 19	11.2	5.4 × 2.3	Sc	

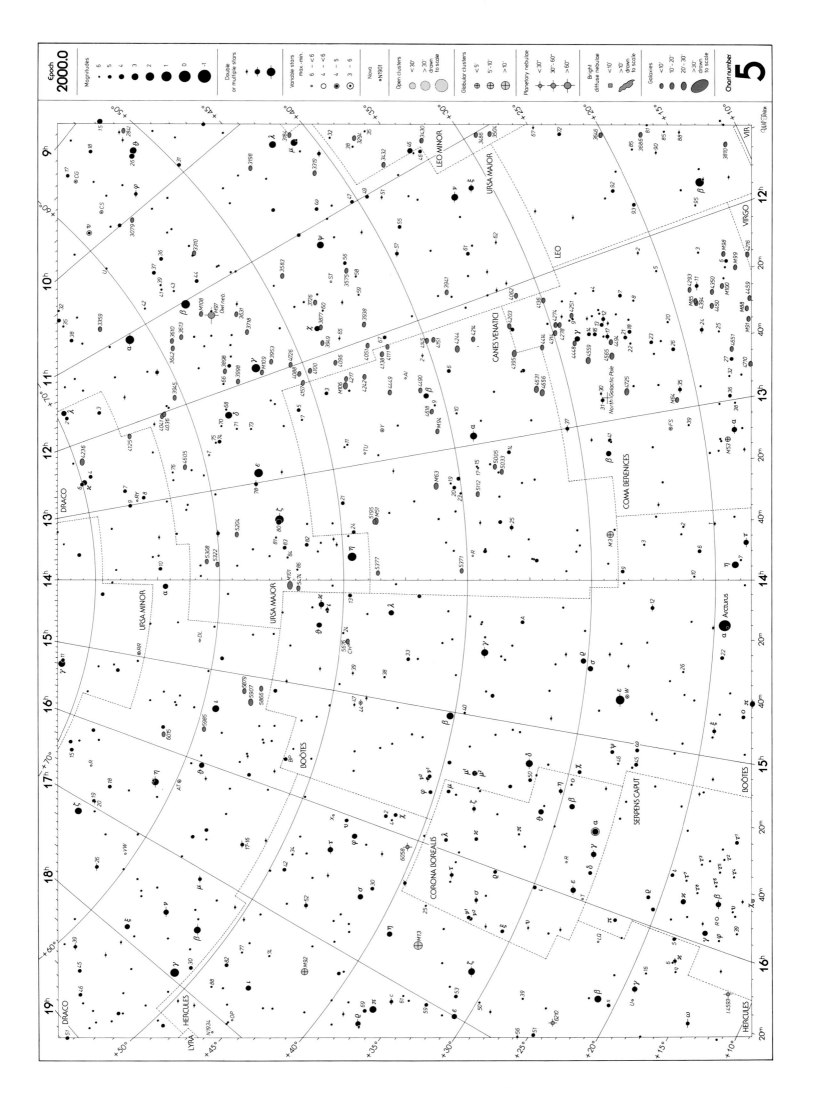

Chart 6 RA 16h to 20h. Dec +20° to +70°

Variable stars

		RA h m	Dec °	Range	Type	Period d	Spectrum
χ	Cyg	19 50.6	+32 55	3.3–14.2	M	406.9	S
R	Cyg	19 36.8	+50 12	6.1–14.2	M	426.4	M
AF	Cyg	19 30.2	+46 09	7.4–9.4	SR	94	M
RT	Cyg	19 43.6	+48 47	6.4–12.7	M	190.2	M
R	Dra	16 32.7	+66 45	6.7–13.0	M	245.5	M
AT	Dra	16 17.3	+59 45	6.8–7.5	Irr	—	M
VW	Dra	17 16.5	+60 40	6.0–6.5	SR	170	K
30	Her	16 28.6	+41 53	5.7–7.2	SR	70	M
68	Her	17 17.3	+35 06	4.6–5.3	EB	2.05	B+B
X	Her	16 02.7	+47 14	7.5–8.6	SR	95	M
β	Lyr	18 50.1	+33 22	3.3–4.3	EB	12.94	B+A
R	Lyr	18 55.3	+43 57	3.9–5.0	SR	46	M
XY	Lyr	18 38.1	+39 40	7.3–7.8	Irr	—	M
U	Vul	19 36.6	+20 20	6.8–7.5	Cep	7.99	F-G

Double stars

		RA h m	Dec °	PA	Sep	Mag
β	Cyg	19 30.7	+27 58	054	34.4	3.1, 5.1 Albireo
δ	Cyg	19 45.0	+10 46	225	2.4	2.9, 6.3 Binary, 828 y
η	Cyg	19 56.3	+35 05	AB 208	7.4	3.9,11.9
				AC 327	46.0	10.4
				AD 169	49.7	10.4
				AE 246	60.2	11.4
ψ	Cyg	19 55.6	+52 26	178	3.2	4.9, 7.4
17	Cyg	19 46.4	+33 44	069	26.0	5.0, 9.2
β	Dra	17 30.4	+52 18	AB 013	4.2	2.8,13.8
				AC 156	117.4	12.5
η	Dra	16 24.0	+61 31	142	5.2	2.7, 8.7
μ	Dra	17 05.3	+54 28	020	1.9	5.7, 5.7 Binary 482 y
ν	Dra	17 32.2	+55 11	312	61.9	4.9, 4.9
o	Dra	18 51.2	+59 23	326	34.2	4.8, 7.8
ϖ	Dra	17 36.9	+68 45	276	72.3	4.9,13.2
17	Dra	16 36.2	+52 55	AB 108	3.4	5.4, 6.4 } C is 16 Dra
				AC 194	90.3	5.5
39	Dra	18 23.9	+58 48	AB 351	3.8	5.0, 8.0
				AC 021	88.9	7.4
				AE 066	198.0	10.9
				AF 080	150.3	11.2
				AG 254	36.0	14.2
				AH 086	41.3	14.2
σ	Her	17 15.0	+24 30	236	8.9	3.7, 8.2 Optical
ζ	Her	16 41.3	+31 36	089	1.6	2.9, 5.5 Binary, 34.5 y
μ	Her	17 46.5	+27 43	247	33.8	3.4,10.1
ρ	Her	17 23.7	+37 09	316	4.1	4.6, 5.6
τ	Her	16 19.7	+46 19	146	6.7	3.9,14.6
68	Her	17 17.3	+33 06	060	4.4	4.8v,10.2
90	Her	17 53.3	+40 00	116	1.6	5.2, 8.5
95	Her	18 01.5	+21 36	258	6.3	5.0, 5.1
100	Her	18 07.8	+26 06	183	14.2	5.9, 6.0
102	Her	18 08.8	+20 49	136	23.4	4.4,11.9
δ^1	Lyr	18 53.7	+36 58	020	174.6	5.6, 9.3
δ^2	Lyr	18 54.5	+36 54	349	86.2	4.5,11.2
ε	Lyr	18 44.3	+39 40	AB+CD 173	207.7	4.7, 5.1
				AB 357	2.6	5.0, 6.1
				CD 094	2.3	5.2, 5.5
ξ	Lyr	18 44.8	+37 36	150	43.7	4.3, 5.9
η	Lyr	19 13.8	+39 09	082	28.1	4.4, 9.1
α-8	Vul	19 28.7	+24 40	028	413.7	4.4, 5.8
2	Vul	19 17.7	+23 02	127	1.8	5.4, 9.2

Open clusters

NGC		RA	Dec	Diam	Mag	N*
6791	Lyr	19 20.7	+37 51	16	9.5	300
6811	Cyg	19 38.2	+46 34	13	6.8	70
6819	Cyg	19 41.3	+40 11	5	7.3	—
6823	Vul	19 43.1	+23 18	12	7.1	30
6830	Vul	19 51.0	+23 04	12	7.9	20
6834	Cyg	19 52.2	+29 25	5	7.8	50

Globular cluster

NGC		RA	Dec	Diam	Mag	
6205	Her	16 41.7	+36 28	16.6	5.9	M13
6341	Her	17 17.1	+43 08	11.2	6.5	M92
6779	Lyr	19 16.6	+30 11	7.1	8.2	M56

Planetary nebulae

NGC		RA	Dec	Diam	Mag	Mag*	
6058	Her	16 04.4	+40 41	23	13.3	13.8	
6210	Her	16 44.5	+23 49	14	9.3	12.9	
6543	Dra	17 58.6	+66 38	18×350	8.8	11.4	
6720	Lyr	18 53.6	+33 02	70×150	9.7	14.8	M57 Ring Nebula
6826	Cyg	19 44.8	+50 31	30×140	9.8	10.4	Blinking Nebula
6853	Vul	19 59.6	+22 43	350×910	7.6	13.9	M27 Dumbbell Nebula

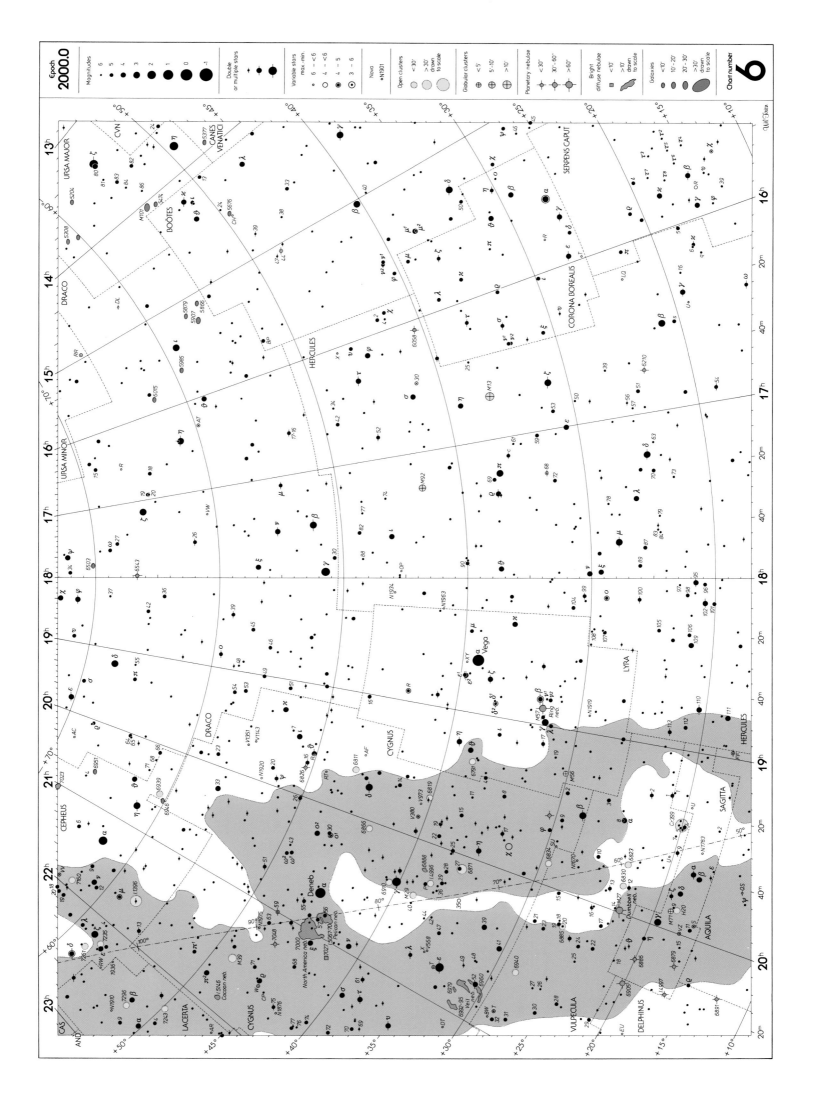

Chart 7 RA 20h to 0h. Dec +20° to +70°

Variable stars

		RA h m	Dec °	Range	Type	Period d	Spectrum
ρ	Cas	23 53.4	+58 30	4.1– 6.2	?	—	F–K
R	Cas	23 58.4	+51 24	4.7–13.5	M	430.5	M
δ	Cep	22 29.2	+58 25	3.5– 4.4	Cep	5.37	F–G
T	Cep	21 09.5	+68 29	5.2–11.3	M	388.1	M
RW	Cep	22 23.1	+55 58	8.6–10.7	SR	346	K
VV	Cep	21 56.7	+63 38	4.8– 5.4	E	7430	M+B
μ	Cep	21 43.5	+58 47	3.4– 5.1	SR?	730	M
P	Cyg	20 17.8	+38 02	3.0– 6.0	Irr	—	Bp
T	Cyg	20 47.2	+34 22	5.0– 5.5	Irr	—	K
X	Cyg	20 43.4	+35 35	5.9– 6.9	Cep	16.39	F–G
β	Peg	23 03.8	+28 05	2.3– 2.7	SR	38	M

Double stars

		RA	Dec	PA	Sep	Mag	
κ	And	23 40.4	+44 20	AB 194	46.8	4.1,11.1	
				AC 294	107.0	11.1	
2	And	23 02.6	+42 45	AB 345	0.4	5.1, 8.8 Binary, 77 y	
				AC 192	90.4	13.7	
σ	Cas	23 59.0	+55 45	326	3.0	5.0, 7.1	
6	Cas	23 48.8	+62 13	193	1.6	5.5, 8.0	
δ	Cep	22 29.2	+58 25	191	41.0	var, 7.5	
η	Cep	20 45.3	+61 50	066	51.7	3.4,11.1 Optical	
ξ	Cep	22 03.8	+64 38	277	7.7	4.4, 6.5 Binary	
o	Cep	22 18.6	+68 07	220	2.9	4.9, 7.1 Binary, 796 y	
ε	Cyg	20 46.2	+33 58	272	54.9	2.5,11.5	
τ	Cyg	21 14.8	+38 03	015	0.5	3.8, 6.4 Binary, 50 y	
μ	Cyg	21 44.1	+28 45	300	1.6	4.8, 6.1 Binary, 507 y	
ν	Cyg	21 17.9	+38 54	AB 220	15.1	4.4,10.0	
				AC 181	21.5	10.0	
69	Cyg	21 25.8	+36 40	AB 030	33.0	5.9,10.3	
				AC 098	54.0	9.0	
8	Lac	22 35.9	+39 38	AB 186	22.4	5.7, 6.5	
				AC 169	48.8	10.5	
				AD 144	81.8	9.3	
				AE 239	336.8	7.8	
κ	Peg	21 44.6	+25 39	095	0.3	4.7, 5.0 Binary, 11.6 y	
2	Peg	21 29.9	+23 38	332	29.8	4.6,11.6	
32	Peg	22 21.3	+28 20	AB 127	72.6	4.8, 9.1	
				AD 307	42.3	11.8	
				AE 116	60.3	11.8	
				BC 018	2.4	10.8	
72	Peg	23 34.0	+31 20	085	0.5	5.7, 5.8 Binary, 241 y	
78	Peg	23 44.0	+29 22	235	1.0	5.0, 8.1	
16	Vul	20 02.0	+24 56	115	0.8	5.8, 6.2	

Open clusters

NGC		RA	Dec	Diam	Mag	N*	
6871	Cyg	20 05.9	+35 47	20	5.2	15	
6866	Cyg	20 03.7	+44 10	7	7.6	80	
6885	Vul	20 12.0	+26 29	7	5.7	30	
6910	Cyg	20 23.1	+40 47	8	7.4	50	
6913	Cyg	20 23.9	+38 32	7	6.6	50	M29
6939	Cyg	20 31.4	+60 38	8	7.8	80	
6940	Vul	20 34.6	+28 18	31	6.3	60	
7067	Cyg	21 24.2	+48 01	3	9.7	20	
7092	Cyg	21 32.2	+48 26	32	4.6	30	M39
7160	Cep	21 53.7	+62 36	7	6.1	12	
7235	Cep	22 12.6	+57 17	4	7.7	30	
7243	Lac	22 15.3	+49 53	21	6.4	40	
7261	Cep	22 20.4	+58 05	6	8.4	30	
7296	Lac	22 28.2	+52 17	4	9.7	20	
7510	Cep	23 11.5	+60 34	4	7.9	60	
7654	Cas	23 24.2	+61 35	13	6.9	100	M52
7686	And	23 30.2	+49 08	15	5.6	20	
7788	Cas	23 56.7	+61 24	9	9.4	20	
7789	Cas	23 57.0	+56 44	16	6.7	300	
7790	Cas	23 58.4	+61 13	17	8.5	40	
IC 1396	Cep	21 39.1	+57 30	50	3.5	50	
H.21	Cas	23 54.1	+61 46	4	9.0	6	

Globular cluster

NGC		RA	Dec	Diam	Mag	N*	
7078	Peg	21 30.0	+12 10	12.3	6.3	M15	

Planetary nebulae

NGC		RA	Dec	Diam	Mag	Mag*	
7048	Cyg	21 14.2	+46 16	61	11.3	18	
7662	And	23 25.9	+42 33	20×130	9.2	13.2	

Nebulae

NGC		RA	Dec	Diam	Mag*	
6960	Cyg	20 45.7	+30 43	70× 6	—	Filamentary Nebula: 52 Cyg
6992/5	Cyg	20 56.4	+31 43	60× 8	—	Veil Nebula: SNR
7000	Cyg	20 58.8	+44 20	120×100	6	North America Nebula
7023	Cep	21 01.8	+68 12	18× 18	6.8	
7635	Cas	23 20.7	+61 12	15× 8	7	Bubble Nebula
IC 5067/70	Cyg	20 50.8	+44 21	80× 70	—	Pelican Nebula
IC 5146	Cyg	21 53.5	+47 16	12× 12	10	Cocoon Nebula, with sparse cluster

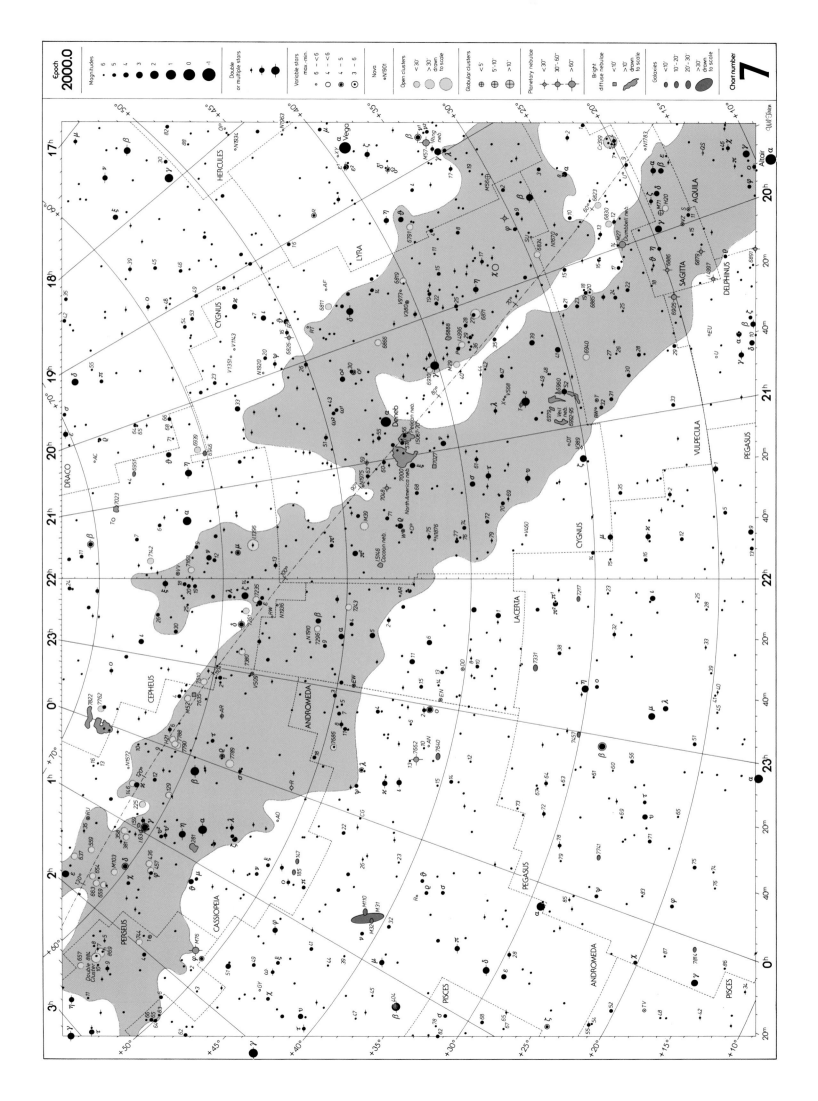

Chart 8 *RA 0h to 4h. Dec +20° to −20°*

Variable stars

		RA h	m	Dec °	Range	Type	Period d	Spectrum
U	Ari	03	11.0	+14 48	7.2–15.2	M	371.1	M
RZ	Ari	02	55.8	+18 20	5.6– 6.1	SR	30	M
o	Cet	02	19.3	−02 59	1.7–10.1	M	332.0	M Mira
Z	Eri	02	47.9	−12 28	7.0– 8.6	SR	80	M
TV	Psc	00	28.0	+17 54	4.6– 5.4	SR	70	M

Double stars

		RA	Dec	PA	Sep	Mag	
γ	Ari	01 53.5	+19 18	000	7.8	4.8, 4.8	
π	Ari	02 49.3	+17 28	{AB 120 / AC 110}	3.2 / 25.2	5.2, 8.7 / 10.8	
γ	Cet	02 43.3	+03 14	294	2.8	3.5, 7.3	
ε	Cet	02 39.6	−11 52	039	0.1	5.8, 5.8 Binary, 2.7 y	
θ	Cet	01 24.0	−08 11	057	65.4	3.6,14.6	
ν	Cet	02 35.9	+05 36	081	8.1	4.9, 9.5	
o	Cet	02 19.3	−02 59	085	0.3	var,12.0 Binary, 400 y	
χ	Cet	01 49.6	−10 41	250	183.8	4.9, 6.9	
12	Cet	00 30.0	−03 57	194	10.3	5.7,10.5	
13	Cet	00 35.2	−03 36	224	0.2	5.6, 6.3 Binary, 6.9 y	
26	Cet	01 03.8	+01 22	253	16.0	6.2, 8.6	
37	Cet	01 14.4	−07 55	331	49.7	5.2, 8.7	
66	Cet	02 12.8	−02 24	{AB 234 / AC 061}	16.5 / 172.7	5.7, 7.5 / 11.4	
84	Cet	02 41.2	−00 42	310	4.0	5.8, 9.0	
95	Cet	03 18.4	−00 56	245	1.2	5.6, 7.5 Binary, 217 y	
ρ²	Eri	03 02.7	−07 41	075	1.8	5.3, 9.5	
γ	Peg	00 13.2	+15 11	285	63.4	2.8,11.7	
α	Psc	02 02.0	+02 46	279	1.9	4.2, 5.1 Binary, 933 y	
ξ	Psc	01 13.7	+07 35	063	23.0	5.6, 6.5	
η	Psc	01 31.5	+15 21	036	1.0	3.6,10.6	
ψ¹	Psc	01 05.6	+21 28	{AB 159 / AC 123}	30.0 / 92.6	5.6, 5.8 / 11.2	
51	Psc	00 32.4	+06 57	083	27.5	5.7, 9.5	

Planetary nebula

NGC		RA	Dec	Diam	Mag*
246	Cet	00 47.0	−11.53	225	11.9

Galaxies

NGC		RA	Dec	Mag	Diam	Type	
428	Cet	01 12.9	+00 59	11.3	4.1×3.2	Sc	
470	Psc	01 19.7	+03 25	11.9	3.0×2.0	Sc	
474	Psc	01 20.1	+03 25	11.1	7.9×7.2	S0	
488	Psc	01 21.8	+05 15	10.3	5.2×4.1	Sb	
524	Psc	01 24.8	+09 32	10.6	3.2×3.2	E1	
584	Cst	01 31.3	−06 52	10.3	3.8×2.4	E4	
628	Psc	01 36.7	+15 47	9.2	10.2×9.5	Sc	M74
720	Cet	01 53.0	−13 44	10.2	4.4×2.8	E3	
772	Ari	01 59.3	+19 01	10.3	7.1×4.5	Sb	Arp 78
864	Cet	02 15.5	+06 00	11.0	4.6×3.5	Sc	
895	Cet	02 21.6	−05 31	11.8	3.6×2.8	Sb	
936	Cet	02 27.6	−01 09	10.1	5.2×4.4	SBa	
1042	Cet	02 40.4	−08 26	10.9	4.7×3.9	Sc	
1055	Cet	02 41.8	+00 26	10.6	7.6×3.0	Sb	
1073	Cet	02 43.7	+01 23	11.0	4.9×4.6	SBc	
1084	Eri	02 46.0	−07 35	10.6	2.9×1.5	Sc	
1087	Cet	02 46.4	−00 30	11.0	3.5×2.3	Sc	
1179	Eri	03 02.6	−18 54	11.8	4.6×3.9	Sp	
1300	Eri	03 19.7	−19 25	10.4	6.5×4.3	SBp	
1337	Eri	03 28.1	−08 23	11.7	6.8×2.0	S	
1407	Eri	03 40.2	−18 35	9.8	2.5×2.5	E0	
1068	Cet	02 42.7	−00 01	8.8	6.9×5.9	SBp	M77
7814	Peg	00 03.3	+16 09	10.5	6.3×2.6	Sb	

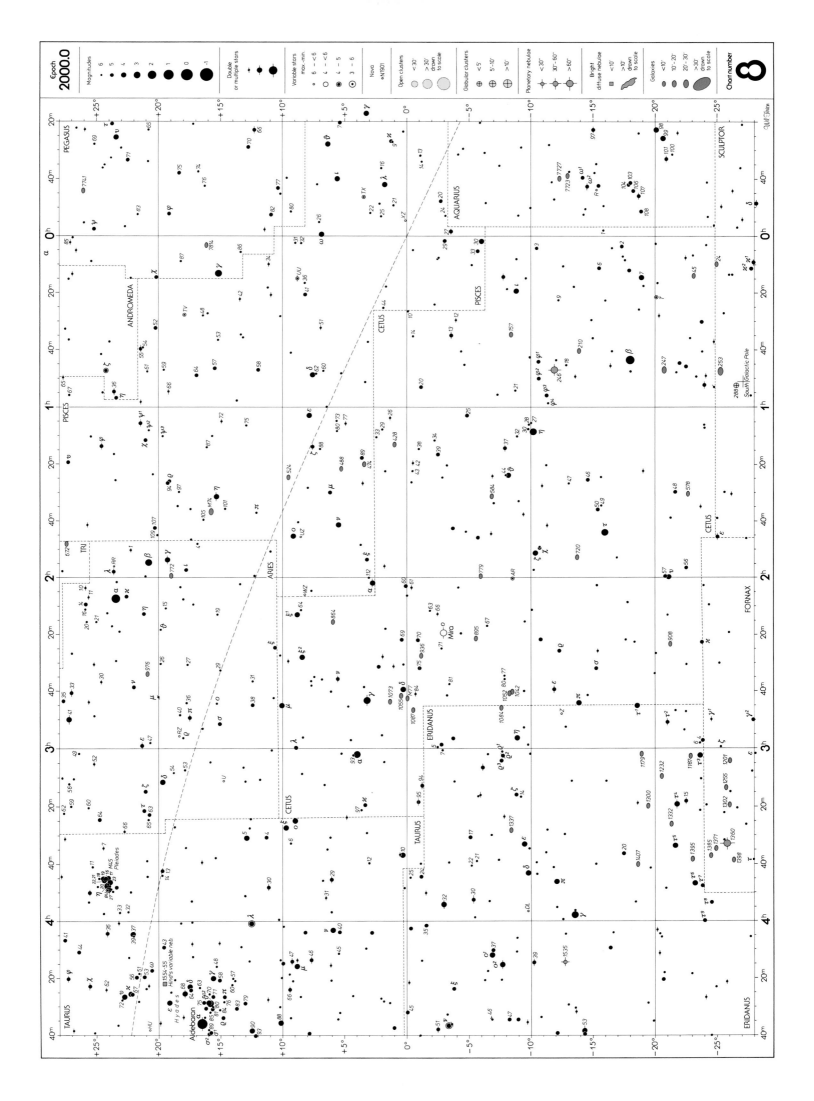

Chart 9 RA 4h to 8h. Dec +20° to −20°

Variable stars

Name	Con	RA h m	Dec	Range	Type	Period d	Spectrum	Notes
R	CMa	07 19.5	−16 24	5.7–6.3	EA	1.14	F	
R	Lep	04 59.6	−14 48	5.5–11.7	M	432.1	N	
RX	Lep	05 11.4	−11 51	5.0–7.0	Irr	—	M	
U	Mon	07 30.8	−09 47	6.1–8.1	RV Tau	92.3	F–K	
V	Mon	06 22.7	−02 12	6.0–13.7	M	333.8	M	
α	Ori	05 55.2	+07 24	0.1–0.9	SR	2110	M	
VV	Ori	05 33.5	−01 09	5.1–5.5	EB	1.49	B+A	
W	Ori	05 05.4	+01 11	5.9–7.7	SR	212	N	
BL	Ori	06 25.5	+14 43	8.5–9.7	Irr	—	N	
CK	Ori	05 30.3	+04 12	5.9–7.1	SR	120	K	
λ	Tau	04 00.7	+12 29	3.3–3.8	EA	3.95	B+A	
CE	Tau	05 32.2	+18 36	6.1–6.5	SR	165	M	

Double stars

Name	Con	RA h m	Dec	PA	Sep	Mag	Notes
α	CMa	06 45.1	−16 43	005	4.5	1.5, 8.5	Sirius
μ	CMa	06 56.1	−14 03	AB 340 / AC 288 / AD 061	3.0 / 88.4 / 101.3	5.3, 8.6 / 10.5 / 10.7	
ν¹	CMa	06 36.4	−18 40	262	17.5	5.8, 8.5	
α	CMi	07 39.3	+05 14	021	5.2	0.4,12.9	
η	CMi	07 28.0	+06 57	025	4.0	5.3,11.1	
ο²	Eri	04 15.2	−07 39	104	83.4	4.4, 9.5	
53	Eri	04 38.2	−14 18	029	0.7	4.0, 7.0	
54	Eri	04 40.4	−19 40	161	0.3	4.9, 5.2	
55	Eri	04 43.6	−08 48	317	9.2	6.7, 6.8	
66	Eri	05 06.8	−04 39	009	52.8	5.2, 8.4	
λ	Gem	07 18.1	+16 32	033	9.6	3.6,10,7	
38	Gem	06 54.6	+13 11	147	7.0	4.7, 7.7	Binary, 3190 y
α	Lep	05 32.7	−17 49	AB 156 / AC 186	35.8 / 91.4	2.6,11.1 / 11.8	
τ	Lep	05 12.3	−11 52	337	12.7	4.5,10.8	
κ	Lep	05 13.2	−12 56	358	2.6	4.5, 7.4	
β	Mon	06 28.8	−07 02	AB 132 / AC 124 / AD 056	7.3 / 10 / 25.9	4.7, 5.2 / 6.1 / 12.2	
γ	Mon	06 14.9	−06 16	027	51.4	4.0,13.0	
ε	Mon	06 23.8	+04 36	027	13.4	4.5, 6.5	
S(15)	Mon	06 41.0	+09 54	AB 213 / AC 013 / AC 051 / AD 308 / AE 139 / AF 222 / AK 056	2.8 / 16.6 / 41.3 / 73.9 / 156.0 / 105.6	4.7, 7.5 / 9.8 / 9.6 / 9.9 / 7.7 / 8.1	
β	Ori	05 14.5	−08 12	202	9.5	0.1, 6.8	Rigel
ξ	Ori	05 40.8	−01 57	AB 162 / AC 010	2.4 / 57.6	1.9, 4.0 / 9.9	Binary, 1509 y
η	Ori	05 24.5	−02 24	AB 080 / AC 051	1.5 / 115.1	3.8, 4.8 / 9.4	
θ	Ori	05 35.3	−05 23	AB 031 / AC 132 / AD 096	8.8 / 12.8 / 21.5	6.7, 7.9 / 5.1 / 6.7	Trapezium
ι	Ori	05 35.4	−05 55	141	11.3	2.8, 6.9	
μ	Ori	06 02.4	+09 39	AB 023	0.4	4.4, 6.0	
π³	Ori	04 49.8	+06 58	138	94.6	3.2, 8.7	
ρ	Ori	05 13.3	+02 54	064	7.0	4.5, 8.3	
14	Ori	05 07.9	−08 30	349	0.7	5.8, 6.5	Binary, 199 y
75	Ori	06 17.1	+09 57	AB 258 / AB 159	62.7 / 117.3	5.4, 9.5 / 8.5	
α	Tau	04 35.9	+16 31	AB 110 / AC 034	30.4 / 121.7	0.9,13.4 / 11.1	Optical
θ	Tau	04 28.7	+15 32	346	337.4	3.4, 3.8	
σ	Tau	04 39.3	+15 55	193	431.2	4.7, 5.1	
47	Tau	04 13.9	+09 16	351	1.1	4.9, 7.4	
66	Tau	04 23.9	+09 28	265	0.1	5.8, 5.9	Binary, 51.6 y
126	Tau	05 41.3	+16 32	238	0.3	5.3, 5.9	

Open clusters

NGC	Con	RA h m	Dec	Diam	Mag	N*	Notes
— .	Tau	04 27	+16	330	1	200+	Hyades
1647	Tau	04 46.0	+19 04	45	6.4	200	
1807	Tau	05 10.7	+16 32	17	7.0	20	Asterism?
1817	Tau	05 12.1	+16 42	16	7.7	60	
1981	Ori	05 35.2	−04 26	25	4.6	20	
2112	Ori	05 53.9	+00 24	11	9.1	50	
2169	Gem	06 08.4	+13 57	7	5.9	30	
2186	Gem	06 12.2	+05 27	4	8.7	30	
2215	Mon	06 21.0	−07 17	11	8.4	40	
2244	Mon	06 32.4	+04 52	24	4.8	100	In Rosette Nebula
2251	Mon	06 34.7	+08 22	10	7.3	30	
2286	Mon	06 47.6	−03 10	15	7.5	50	
2301	Mon	06 51.8	+00 28	12	6.0	80	
2323	Mon	07 03.2	−08 20	16	5.9	80	M50
2335	Mon	07 06.6	−10 05	12	7.2	35	
2343	Mon	07 08.3	−10 39	7	6.7	20	
2345	CMa	07 08.3	−13 10	12	7.7	20	
2353	Mon	07 14.6	−10 18	20	7.1	30	
2355	Gem	07 16.9	+13 47	9	9.7	40	
2360	CMa	07 17.8	−15 37	13	7.2	80	Asterism?
2395	Gem	07 27.1	+13 35	12	8.0	30	
2422	Pup	07 36.6	−14 30	30	4.4	30	M47
2437	Pup	07 41.8	−14 49	27	6.1	100	M46 Contains p.n. NGC
2479	Pup	07 55.1	−17 43	7	9.6	45	
Hel 71	Pup	07 37.5	−12 04	9	7.1	80	

Planetary nebulae

NGC	Con	RA h m	Dec	Diam	Mag	Mag*	Notes
1535	Eri	05 14.2	−12 44	18×44	9.6	12.2	
2438	Pup	07 41.8	−14 44	66	10.1	17.7	In cluster NGC 2437
IC 418	Lep	05 27.5	−12 42	12	10.7	10.7	

Nebulae

NGC	Con	RA h m	Dec	Diam	Mag*	Notes
1554-5	Tau	04 21.8	+19 32	var	9v	Hind's Variable Nebula (*T* Tau)
1976	Ori	05 35.4	−05 27	66×60	5	M42
1982	Ori	05 35.6	−05 16	20×15	7	M43 Extension of M42
2068	Ori	05 46.7	+00 03	8× 6	10	M78 Nebula is mag 8
2149	Mon	06 03.5	−09 44	3× 2	9	
2237-9	Mon	06 32.3	+05 03	80×60	—	Rosetta Nebula. Contains cluster NGC 2244
2261	Mon	06 39.2	+08 44	2× 1	10v	R Mon: Hubble's Variable Nebula
2264	Mon	06 40.9	+09 54	60×30	4v	S Mon: Cone Nebula
I.C. 434	Ori	05 41.0	−02 24	60×10	2(ζ)	Behind Horse's Head Nebula. Barnard 33

Galaxies

NGC	Con	RA h m	Dec	Mag	Diam	Type
1637	Eri	04 41.5	−02 51	10.9	3.3×2.9	Sc

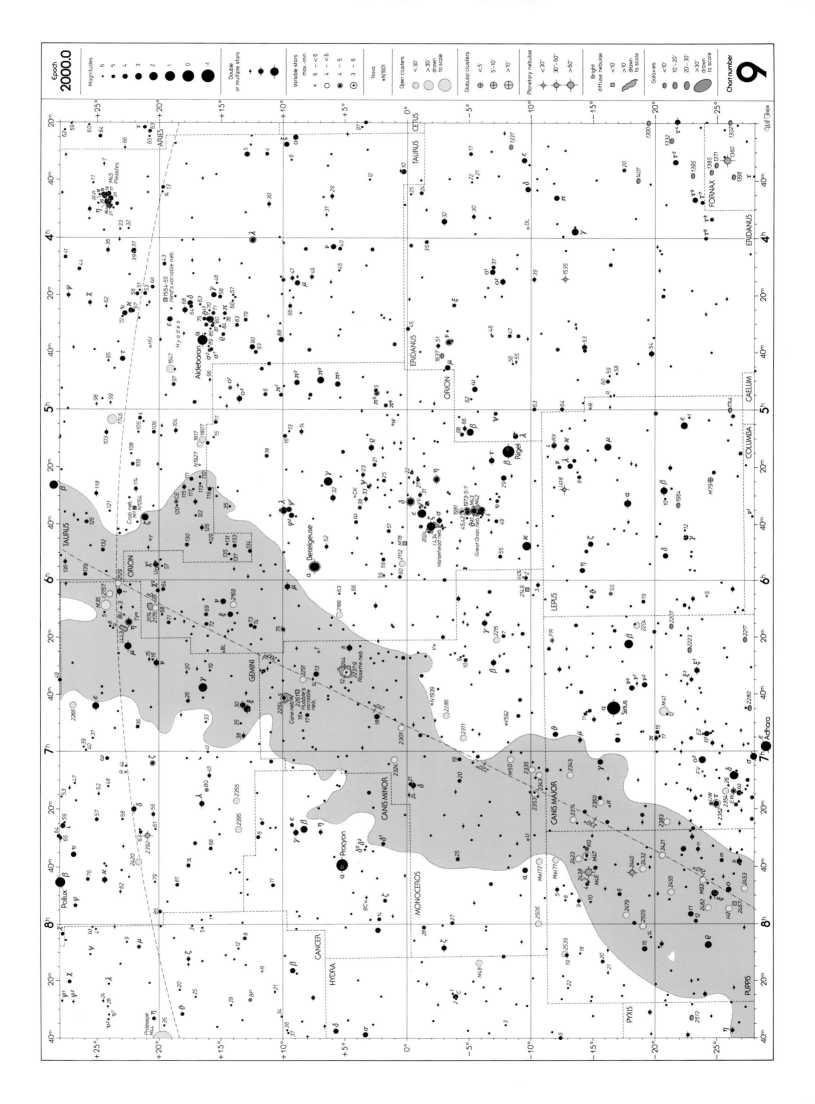

Chart 10 RA 8h to 12h. Dec +20° to −20°

Variable stars

		RA h	m	Dec °		Range	Type	Period d	Spectrum
R	Cnc	08	16.6	+11 44		6.1–11.8	M	361.6	M
X	Cnc	08	55.4	+17 14		5.6– 7.5	SR	195	N
IN	Hya	09	20.6	+00 11		6.3– 6.7	SR	45	M
R	Leo	09	47.6	+11 25		4.4–11.3	M	312.4	M
VY	Leo	10	56.0	+06 11		5.7– 6.0	Irr	—	M

Double stars

		RA h	m	Dec °	PA	Sep	Mag
δ	Cnc	08	44.7	+18 09	090	38.4	3.9,11.9 Optical
ζ	Cnc	08	12.2	+17 39	AB+C 088	5.7	5.0, 6.2 Binary, 1150 y
					AB 182	0.6	5.3, 6.0 Binary, 59.7 y
					AB+D 108	287.9	9.7
γ	Crt	11	24.9	−17 41	096	5.2	4.1, 9.6
ι	Crt	11	38.7	−13 12	226	1.4	5.5,10.9
α	Hya	09	27.6	−08 40	153	283.1	2.0, 9.5
ε	Hya	08	46.8	+06 25	AB 295	0.2	3.8, 4.7 Binary, 890 y
					AB+C 281	2.8	6.8
θ	Hya	09	14.4	+02 19	197	29.4	3.9, 9.9
27	Hya	09	20.5	−09 33	211	229.4	5.0, 6.9
29	Hya	09	27.2	−09 13	AB 184	0.2	7.2, 7.3
					AB+C 175	10.8	11.8
α	Leo	10	08.4	+11 58	307	176.9	1.4, 7.7 Regulus
γ	Leo	10	20.0	+19 51	AB 124	4.3	2.2, 3.5 Binary, 619 y
					AC 291	259.9	9.2
					AD 302	333.0	9.6
ι	Leo	11	23.9	+10 32	131	1.5	4.0, 6.7 Binary, 192 y
ϖ	Leo	09	28.5	+09 03	053	0.5	5.9, 6.5 Binary, 118 y
χ	Leo	11	05.0	+07 20	AB 262	3.3	4.6,10.9
					AC 305	276.4	8.9
TX	Leo	10	35.0	+08 39	157	2.4	5.8, 8.5
3	Leo	09	28.5	+08 11	080	25.2	5.7,10.4
6	Leo	09	32.0	+09 43	075	37.4	5.2, 8.2
31	Leo	10	07.9	+10 00	044	7.9	4.4,13.4
34	Leo	10	11.6	+13 21	286	0.2	6.8, 7.6
90	Leo	11	34.7	+16 48	AB 209	3.3	6.0, 7.3
					AC 234	63.1	8.7
γ	Sex	09	52.5	−08 06	AB 067	0.6	5.6, 6.1 Binary, 75.6 y
					AC 325	35.8	12.0

Open clusters

NGC		RA	Dec	Diam	Mag	N*	
2506	Mon	08 00.2	−10 47	7	7.6	150	
2509	Pup	08 00.7	−19 04	8	9.3	70	
2539	Pup	08 10.7	−12 50	22	6.5	50	
2548	Hya	08 13.8	−05 48	54	5.8	80	M48
2632	Cnc	08 40.1	+19 59	95	3.1	50	M44 Praesepe
2682	Cnc	08 50.4	+11 49	30	6.9	200	M67

Planetary nebula

NGC		RA	Dec	Diam	Mag	Mag*	
3242	Hya	10 24.8	−18 38	16×1250	8.6	12.0	Ghost of Jupiter

Galaxies

NGC		RA	Dec	Mag	Diam	Type	
2775	Cnc	09 10.3	+07 02	10.3	4.5×3.5	Sa	
2967	Sex	09 42.1	+00 20	11.6	3.0×2.9	Sc	
3115	Sex	10 05.2	−07 43	9.1	8.3×3.2	E6	
3166	Sex	10 13.8	+03 26	10.6	5.2×2.7	SBa	
3169	Sex	10 14.2	+03 28	10.4	4.8×3.2	Sb	
3351	Leo	10 44.0	+11 42	9.7	7.4×5.1	SBb	M95
3368	Leo	10 46.8	+11 49	9.2	7.1×5.1	Sb	M96
3377	Leo	10 47.7	+11 59	10.2	4.4×2.7	E5	
3379	Leo	10 47.8	+12 35	9.3	4.5×4.0	E1	M105
3384	Leo	10 48.3	+12 38	10.0	5.9×2.6	E7	
3412	Leo	10 50.9	+13 25	10.6	3.6×2.0	E5	
3489	Leo	11 00.3	+13 54	10.3	3.7×2.1	E6	
3521	Leo	11 05.8	−00 02	8.9	9.5×5.0	Sb	
3571	Crt	11 11.5	−18 17	12.8	3.3×1.3	Sa	
3593	Leo	11 14.6	+12 49	11.0	5.8×2.5	Sb	
3596	Leo	11 15.1	+14 47	11.6	4.2×4.1	Sc	
3607	Leo	11 16.9	+18 03	10.0	3.7×3.2	E1	
3623	Leo	11 18.9	+13 05	9.3	10.0×3.3	Sb	M65
3626	Leo	11 20.1	+18 21	10.9	3.1×2.6	Sb	
3627	Leo	11 20.2	+12 59	9.0	8.7×4.4	Sb	M66
3628	Leo	11 20.3	+13 36	9.5	14.8×3.6	Sb	Arp 317
3630	Leo	11 20.3	+02 58	12.8	2.3×0.9	E7	
3640	Leo	11 21.1	+03 14	10.3	4.1×3.4	E1	
3672	Crt	11 25.0	−09 48	11.5	4.1×2.1	Sb	
3686	Leo	11 27.7	+17 13	11.4	3.3×2.6	Sc	
3810	Leo	11 41.0	+11 28	10.8	4.3×3.1	Sc	
3887	Crt	11 47.1	−16 51	11.0	3.3×2.7	Sc	
3981	Crt	11 56.1	−19 54	12.4	3.9×1.5	Sb	

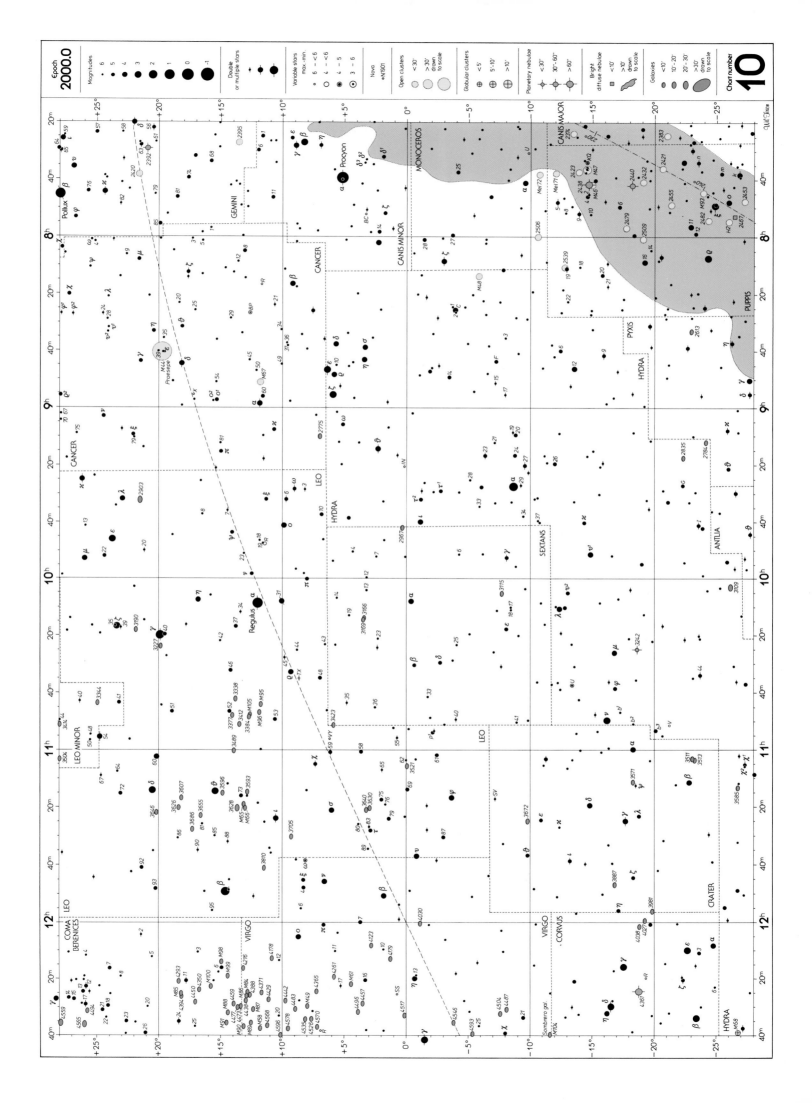

Chart 11 RA 12h to 16h. Dec +20° to −20°

Variable stars

	RA h	m	Dec °	Range	Type	Period d	Spectrum
R Crv	12	19.6	−19 15	6.7–14.4	M	317.0	M
R Ser	15	50.7	+15 08	5.1–14.4	M	356.4	M
δ Lib	15	01.1	−08 31	4.9– 5.9	EA	2.33	B
R Vir	12	38.5	+06 59	6.0–12.1	M	145.6	M
S Vir	13	33.0	−07 12	6.3–13.2	M	377.4	M
SS Vir	12	25.3	+00 48	6.0– 9.6	M	354.7	N

Double stars

	RA		Dec	PA	Sep	Mag	
ζ Boö	14	41.1	+13 44	AB 303	1.0	4.5, 4.6 Binary, 123.3 y	
				AC 259	99.3	10.9	
π Boö	14	40.7	+16 25	108	5.6	4.9, 5.8	
24 Com	12	35.1	+18 23	271	20.3	5.2, 6.7	
δ Crv	12	29.9	−16 31	214	24.2	3.0, 9.2	
α Lib	14	50.9	−16 02	α²-α¹ 314	231.0	2.8, 5.2	
ι Lib	15	12.2	−19 47	111	57.8	5.1, 9.4	
κ Lib	15	41.9	−19 41	279	172.0	4.7, 9.7	
μ Lib	14	49.3	−14 09	AB 355	1.8	5.8, 6.7	
				AC 289	15.0	14.5	
				AD 174	25.0	13.9	
18 Lib	14	58.9	−11 09	AB 039	19.7	5.8,10.0	
				AC 41	162.3	11.3	
47 Lib	15	55.0	−19 23	129	0.5	6.1, 8.1	
β Ser	15	46.2	+15 25	265	30.6	3.7, 9.9	
δ Ser	15	34.8	+10 32	177	4.4	4.1, 5.2 Binary, 3168 y	
5 Ser	15	19.3	+01 46	AB 036	11.2	5.1,10.1	
				AC 040	127.2	9.1	
6 Ser	15	21.0	+00 43	020	3.1	5.4,10.0	
γ Vir	12	41.7	−01 27	287	3.0	3.5, 3.5 Binary, 171.4 y	
θ Vir	13	09.9	−05 32	343	7.1	4.4, 9.4	
τ Vir	14	01.6	+01 33	290	80.0	4.3, 9.6	
17 Vir	12	22.5	+05 18	337	20.0	6.6, 9.4	
31 Vir	12	42.0	+06 48	037	4.0	5.6,11.4	
73 Vir	13	32.0	−18 44	183	0.1	6.7, 6.9	
84 Vir	13	43.1	+03 32	229	2.9	5.5, 7.9	

Open clusters

NGC		RA	Dec	Diam	Mag	N*	
Mel 111	Com	12 25	+26	275	4	80	Coma Berenices

Globular cluster

NGC		RA	Dec	Diam	Mag	N*
5024	Com	13 12.9	+18 10	1	7.7	M53
5904	Ser	15 18.6	+02 05	17	5.8	M5

Planetary nebula

NGC		RA	Dec	Diam	Mag	Mag*
4361	Crv	12 24.5	−18 48	45×110	10.3	13.2

Galaxies*

NGC		RA	Dec	Mag	Diam	Type	
4192	Com	12 13.8	+14 54	10.1	9.5×3.2	Sb	M98
4216	Vir	12 15.9	+13 09	10.0	8.3×2.2	Sb	
4254	Com	12 18.8	+14 25	9.8	5.4×4.8	Sc	M99
4261	Vir	12 19.4	+05 49	10.3	3.9×3.2	E2	
4303	Vir	12 21.9	+04 28	9.7	6.0×5.5	Sc	M61
4321	Com	12 29.9	+15 49	9.4	6.9×6.2	Sc	M100
4374	Vir	12 25.1	+12 53	9.3	5.0×4.4	E1	M84
4429	Vir	12 27.4	+11 07	10.2	5.5×2.6	SO	
4438	Vir	12 27.8	+13 01	10.1	9.3×3.9	Sap	
4442	Vir	12 28.1	+09 48	10.5	4.6×1.9	E5p	
4450	Com	12 28.5	+17 05	10.1	4.8×3.5	Sb	
4406	Vir	12 26.2	+12 57	9.2	7.4×5.5	E3	M86
4459	Com	12 29.0	+13 59	10.4	3.8×2.8	E2	
4472	Vir	12 29.8	+08 00	8.4	8.9×7.4	E4	M49
4473	Com	12 29.8	+13 26	10.2	4.5×2.6	E4	
4477	Com	12 30.0	+13 38	10.4	4.0×3.5	SBa	
4486	Vir	12 30.8	+12 24	8.6	7.2×6.8	E1	M87 Virgo A
4501	Com	12 32.0	+14 25	9.5	6.9×3.9	Sb	M88
4535	Vir	12 34.3	+08 12	9.8	6.8×5.0	SBc	
4546	Vir	12 35.5	−03 48	10.3	3.5×1.7	E6	
4548	Com	12 35.4	+14 30	10.2	5.4×4.4	SBb	M91
4552	Vir	12 35.7	+12 33	9.8	4.2×4.2	E0	M89
4569	Vir	12 36.8	+13 10	9.5	9.5×4.7	Sb	M90
4579	Vir	12 37.7	+11 49	9.8	5.4×4.4	Sb	M58
4594	Vir	12 40.0	−11 37	8.3	8.9×4.1	SBa	M104 Sombrero Hat
4596	Vir	12 39.9	+10 11	10.5	3.9×2.8	SBa	
4621	Vir	12 42.0	+11 39	9.8	5.1×3.4	E3	M59
4636	Vir	12 42.8	+02 41	9.6	6.2×5.6	E1	
4649	Vir	12 43.7	+11 33	8.8	7.2×6.2	E1	M60
4651	Com	12 43.7	+16 24	10.7	3.8×2.7	Scp	
4654	Vir	12 44.0	+13 08	10.5	4.7×3.0	Sc	
4689	Com	12 47.8	+13 46	10.9	4.0×3.5	Sb	
4697	Vir	12 48.6	−05 48	9.3	6.0×3.8	E4	
4699	Vir	12 49.0	−08 40	9.6	3.5×2.7	Sa	
4753	Vir	12 52.4	−01 12	9.9	5.4×2.9	Pec	
4762	Vir	12 52.9	+11 14	10.2	8.7×1.6	SB0	
4856	Vir	12 59.3	−15 02	10.4	4.6×1.6	SBa	
5247	Vir	13 38.1	−17 53	10.5	5.4×4.7	Sb	
5248	Boö	13 37.5	+08 53	10.2	6.5×4.9	Sc	
5363	Vir	13 56.1	+05 15	10.2	4.2×2.7	Ep	
5364	Vir	13 56.2	+05 01	10.4	7.1×5.0	SB+p	

*There are so many galaxies in this region that a selection has had to be made, with a general limiting magnitude of 10.5.

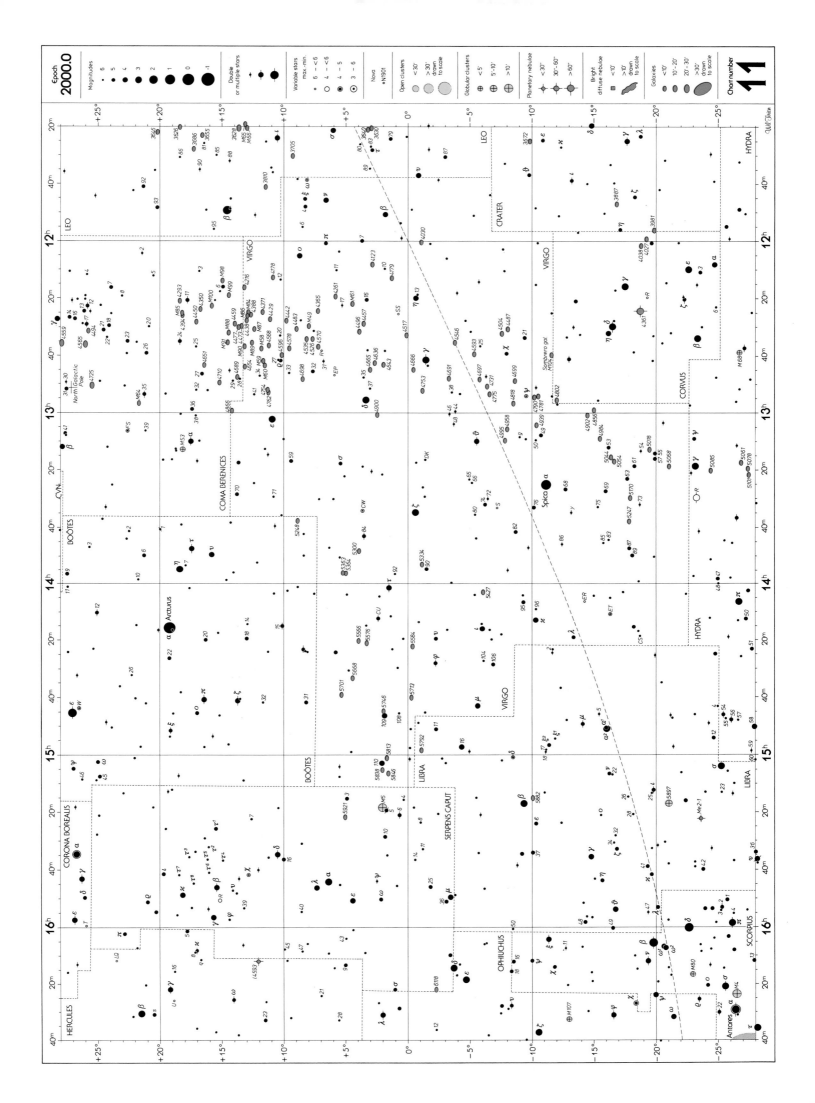

Chart 12 — *RA 16h to 20h. Dec +20° to −20°*

Variable stars

		RA h m	Dec °	Range	Type	Period d	Spectrum
η	Aql	19 52.5	+01 00	3.5– 4.4	Cep	7.18	F-G
R	Aql	19 06.4	+08 14	5.5–12.0	M	284.2	M
U	Aql	19 29.4	−07 03	6.1– 6.9	Cep	7.02	F-G
V	Aql	19 04.4	−05 41	6.6– 8.4	SR	353	N
α	Her	17 14.6	+14 23	3.0– 4.0	SR	~100?	M
S	Her	16 51.9	+14 56	6.4–13.8	M	307.4	M
U	Her	16 25.8	+18 54	6.5–13.4	M	406.0	M
χ	Oph	16 27.0	−18 27	4.2– 5.0	Irr	—	B+B
U	Oph	17 16.5	+01 13	5.9– 6.6	EA	1.68	F-G
Y	Oph	17 52.6	−06 09	5.9– 6.4	Cep	17.12	F-G
R	Sct	18 47.5	−05 42	4.4– 8.2	RV Tau	140	G-K
S	Sge	19 56.0	+16 38	5.3– 6.0	Cep	8.38	F-G
U	Sge	19 18.8	+19 37	6.6– 9.2	EA	3.38	B+K

Double stars

		RA h m	Dec °	PA	Sep	Mag	Notes
α	Aql	19 50.8	+08 52	301	165.2	0.8, 9.5	Altair
γ	Aql	19 46.3	+10 37	238	132.6	2.7,10.7	
δ	Aql	19 25.5	+03 07	271	108.9	3.4,10.9	
ε	Aql	18 59.6	+15 04	187	131.1	4.0, 9.9	
ι	Aql	19 36.7	−01 17	161	47.0	4.4,13.1	
μ	Aql	19 34.1	+07 23	280	31.2	4.5,13.0	
ν	Aql	19 26.5	+00 20	288	201.0	4.7, 8.9	
π	Aql	19 48.7	+11 49	110	1.4	6.1, 6.9	
χ	Aql	19 42.6	+11 50	077	0.5	5.6, 6.8	
U	Aql	19 29.4	−07 03	228	1.5	var,11.7	
23	Aql	19 18.5	+01 05	005	3.1	5.3, 9.3	
31	Aql	19 25.0	+11 57	343	105.6	5.2, 8.7	
57	Aql	19 54.6	−08 14	170	35.7	5.8, 6.5	
α	Her	17 14.6	+14 23	107	4.7	var, 5.4 Binary, 3600 y	
κ	Her	16 08.1	+17 03	012	28.4	5.3, 6.5	
ϖ	Her	16 25.4	+14 02	AB 223 / AC 096	1.0 / 28.4	4.6,11.6 / 11.1	
37	Her	16 40.6	+04 13	230	69.8	5.8, 7.0	
54	Her	16 55.4	+18 26	183	2.5	5.4,12.7	
η	Oph	17 10.4	−15 43	247	0.5	3.0, 3.5 Binary, 84.3 y	
λ	Oph	16 30.9	+01 59	AB 022 / AB+C 170 / AD 246	1.5 / 119.2	4.2, 5.2 Binary, 129.9 y / 11.1	
τ	Oph	18 03.1	−08 11	AB 280 / AC 127	1.8 / 100.3	5.2, 5.9 Binary, 280 y / 9.3	
υ	Oph	16 27.8	−08 22	095	1.0	4.6, 7.8	
φ	Oph	16 31.1	−16 37	037	34.4	4.3,12.8	
X	Oph	18 38.3	+08 50	150	0.4	5.9v,8.6 Binary, 485 y	
19	Oph	16 47.2	+02 04	089	23.4	6.1, 9.4	
41	Oph	17 16.6	−00 27	346	1.0	4.8, 7.8	
53	Oph	17 34.6	+09 35	191	41.2	5.8, 8.5	
70	Oph	18 05.5	+02 30	AB 224	1.5	4.6, 6.0 Binary, 88.1 y 4 faint comps	
73	Oph	18 09.6	+04 00	300	0.4	6.1, 7.0 Binary, 270 y	
β	Sco	16 05.4	−19 48	AC 021	13.6	2.6, 4.9 A is a close double	
ν	Sco	16 12.0	−19 28	AB 003 / AC 337	0.9 / 41.1	4.3, 6.8 Binary, 45.7 y / 6.4	
ξ	Sco	16 04.4	−11 22	AB 040 / AC 051	0.8 / 7.6	4.8, 5.1 / 7.3	
11	Sco	16 07.6	−12 45	257	3.3	5.6, 9.9	
θ	Ser	18 56.2	+04 12	104	22.3	4.5, 4.5	
ν	Ser	17 20.8	−12 51	028	46.3	4.3, 8.3	
59	Ser	18 27.2	+00 12	318	3.8	5.3, 7.6	
α	Sge	19 40.1	+18 01	AB 179 / AC 249	31.5 / 35.8	4.4,13.2 / 14.9	
ζ	Sge	19 49.0	+19 09	AB+C 311 / AB 163	8.6 / 0.3	5.7, 8.7 / 5.5, 6.2 Binary, 22.8 y	

Open clusters

NGC		RA	Dec	Diam	Mag	N*	Notes
M24	SGR	18 16.9	−18 29	90	4.5	–	Not a true cluster. Star-cloud in the Milky Way
6494	Sgr	17 56.8	−19 01	27	5.5	150	M23
6613	Sgr	18 19.9	−17 08	9	6.9	20	M18
6633	Oph	18 27.7	+06 34	27	4.6	30	
6645	Sgr	18 32.6	−16 54	10	8.5	40	
6664	Sct	18 36.7	−08 13	16	7.8	50	EV Scuti cluster
6694	Sct	18 45.2	−09 24	15	8.0	30	
6704	Sct	18 50.9	−05 12	6	9.2	30	
6705	Sct	18 51.1	−06 16	14	5.8	500	M11 Wild Duck
6709	Aql	18 51.5	+10 21	13	6.7	40	
6716	Sgr	18 54.6	−19 53	7	6.9	20	
6755	Aql	19 07.8	+04 14	15	7.5	100	
IC 4665	Oph	17 46.3	+05 43	41	4.2	30	
IC 4725	Sgr	18 31.6	−19 15	32	4.6	31	M25: u Sgr.
H.20	Sgr	19 53.1	+18 20	7	7.7	15	

Open clusters

NGC		RA	Dec	Diam	Mag		
6171	Oph	16 32.5	−13 03	10.0	8.1	M107	
6218	Oph	16 47.2	−01 57	14.5	6.6	M12	
6254	Oph	16 57.1	−04 06	15.1	6.6	M10	
6333	Oph	17 19.2	−18 31	9.3	7.9	M9	
6356	Oph	17 23.6	−17 49	7.2	8.4		
6402	Oph	17 37.6	−03 15	11.7	7.6	M14	
6712	Sct	18 53.1	−08 42	7.2	8.2		
6838	Sge	19 53.8	+18 47	7.2	8.3	M71	

Planetary nebulae

NGC		RA	Dec	Diam	Mag	Mag*	
6572	Oph	18 12.1	+06 51	8	9.0	13.6	
6741	Aql	19 02.6	−00 27	6	10.8	14.7	
6751	Aql	19 05.9	−06 00	20	12.5	13.9	
6790	Aql	19 23.2	+01 31	7	10.2	13.5	
6818	Sgr	19 44.0	−14 09	17	9.9	13.0	
IC 4593	Her	16 12.2	+12 04	12×120	10.9	11.3	

Nebulae

NGC		RA	Dec	Diam	Mag*		
6611	Ser	18 18.8	−13 47	35×28	—	M16	Eagle Nebula, with cluster
6618	Sgr	18 20.8	−16 11	46×37	—	M17	Omega Nebula

Galaxies

NGC		RA	Dec	Diam	Mag	Type	
6118	Ser	16 21.8	−02 17	4.7×2.3	12.3	Sb	
6384	Oph	17 32.4	+07 04	6.0×4.3	10.6	Sb	
6822	Sgr	19 44.9	−14 48	10.2×7.5	9.3	Irr	Barnard's Galaxy

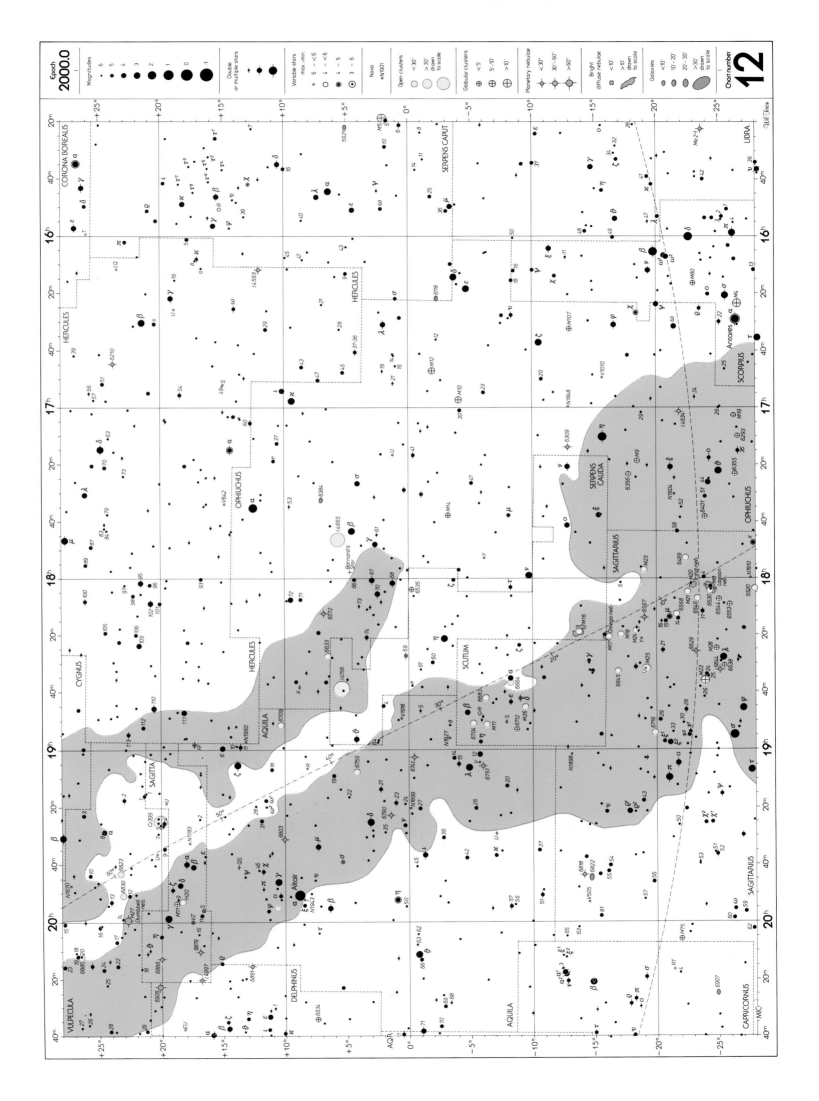

Chart 13 *RA 20h to 0h. Dec +20° to −20°*

Variable stars

		RA h m	Dec °	Range	Type	Period d	Spectrum
R	Aqr	23 43.8	−15 17	5.8-12.4	Symbiotic	387.0	M+Pee
DV	Aqr	20 58.7	−14 29	5.9- 6.2	EB	1.58	F
U	Del	20 45.5	+18 05	7.6- 8.9	SR	110	M
EU	Del	20 37.9	+18 16	5.8- 6.9	SR	59	M
TX	Psc	23 46.4	+03 29	6.9- 7.7	Irr	—	N
XZ	Psc	23 54.8	+00 07	5.5- 6.0	Irr	—	M
VZ	Sge	20 00.1	+17 31	5.3- 5.6	Irr	—	M
AG	Peg	21 51.0	+12 38	6.0- 9.4	Z And	830	WN+M

Double stars

		RA h m	Dec °	PA	Sep	Mag
68	Aql	20 28.4	−03 21	178	9.8	6.1,13.8
β	Aqr	21 31.6	−05 34	AB 321	35.4	2.9,10.8
				AC 186	57.2	11.4
ζ	Aqr	22 28.8	−00 01	200	2.0	4.3, 4.5 Binary, 856 y
ψ¹	Aqr	23 15.9	−09 05	274	80.4	4.3,13.5
ψ³	Aqr	23 19.0	−09 37	174	1.5	5.0,11.0
51	Aqr	22 24.1	−04 50	AB 324	0.5	6.5, 6.5
				AB+D 191	116.0	10.1
				AC 342	54.4	10.2
				AE 133	132.4	8.6
107	Aqr	23 46.0	−18 41	136	6.6	5.7, 6.7
α	Cap	20 18.1	−12 33	α¹-α² 291	377.7	3.6, 4.2
α¹	Cap	20 17.6	−12 30	AB 182	44.3	4.2,13.7
				AC 221	45.4	9.2
α²	Cap	20 18.1	−12 33	AB 172	6.6	3.6,11.0
				AD 156	154.6	9.3
				BC 240	1.2	11.3
ε	Cap	21 37.1	−19 28	047	68.1	4.7, 9.5
π	Cap	20 27.3	−18 13	148	3.2	5.3, 8.9
ρ	Cap	20 28.9	−17 49	158	0.5	5.0,10.0
σ	Cap	20 19.6	−19 07	179	55.9	5.5, 9.0
τ	Cap	20 39.3	−14 57	118	0.3	5.8, 6.3 Binary, 200 y
α	Del	20 39.6	+15 55	224	29.5	3.8,13.3
β	Del	20 37.5	+14 36	167	0.3	4.0, 4.9 Binary, 26.7y
γ	Del	20 46.7	+16 07	268	9.6	4.5, 5.5
κ	Del	20 39.1	+10 05	286	28.8	5.1,11.7
1	Del	20 30.3	+10 54	AB 346	0.9	6.1, 8.1
				AC 349	16.8	14.1
13	Del	20 47.8	+06 00	194	1.6	5.6, 9.2
β	Eql	21 22.9	+06 49	257	34.4	5.2,13.7
γ	Eql	21 10.4	+10 08	AB 268	1.9	4.7,11.5
				AC 005	47.7	12.5
δ	Eql	21 14.5	+10 00	029	0.3	5.2, 5.3 Binary, 5.7 y
ε	Eql	20 59.1	+04 18	AB 285	1.0	6.0, 6.3 Binary, 101.4y
				AB+C 070	10.7	12.4
ε	Peg	21 44.2	+09 52	AD 280	74.8	7.1
ξ	Peg	22 46.7	+12 10	AB 325	81.8	2.4,11.2
				AC 320	142.5	8.4
20	Peg	22 01.1	+13 07	AB 100	11.5	4.2,12.2
				AC 015	145.0	11.0
34	Peg	22 26.6	+04 24	324	54.7	5.7,11.1
35	Peg	22 27.9	+04 02	AB 224	3.5	5.8,12.3
				AC 272	103.3	12.8
37	Peg	22 30.0	+04 26	AB 210	98.3	4.8, 9.8
				AC 241	181.5	9.7
66	Peg	23 23.1	+12 19	118	0.9	5.8, 7.1 Binary, 140 y
2	Psc	22 59.5	+00 58	056	0.1	5.9, 5.9
15	Sgc	20 04.1	+17 04	085	3.8	5.4,13.1
				AB 276	190.7	5.9, 8.1
				AC 320	203.7	6.8

Open cluster

NGC		RA	Dec	Diam	Mag	N*	
6994	Aqr	20 58.9	−12 38	2.8	8.9	4	M73 Asterism

Globular clusters

NGC		RA	Dec	Diam	Mag	
6394	Del	20 34.2	+07 24	5.9	8.9	
6981	Aqr	20 53.5	−12 32	9.2	9.3	M72
7078	Peg	21 30.0	+12 10	12.3	6.3	M15
7089	Aqr	21 33.5	−00 49	12.9	6.5	M2

Planetary nebulae

NGC		RA	Dec	Diam	Mag	
6879	Sge	20 10.5	+16 55	5	13.0	
6891	Del	20 15.2	+12 42	12× 74	11.7	
7009	Aqr	21 04.2	−11 22	25"×100	8.3	Saturn Nebula
IC 4997	Sge	20 20.2	+16 45	2	11.6	

Galaxies

NGC		RA	Dec	Mag	Diam	Type
7479	Peg	23 04.9	+12 19	11.0	4.1×3.2	SBb
7606	Aqr	23 19.1	−08 29	10.8	5.8×2.6	Sb
7723	Aqr	23 38.9	−12 58	11.1	3.6×2.6	Sb
7727	Aqr	23 39.9	−12 18	10.7	4.2×3.4	SBap

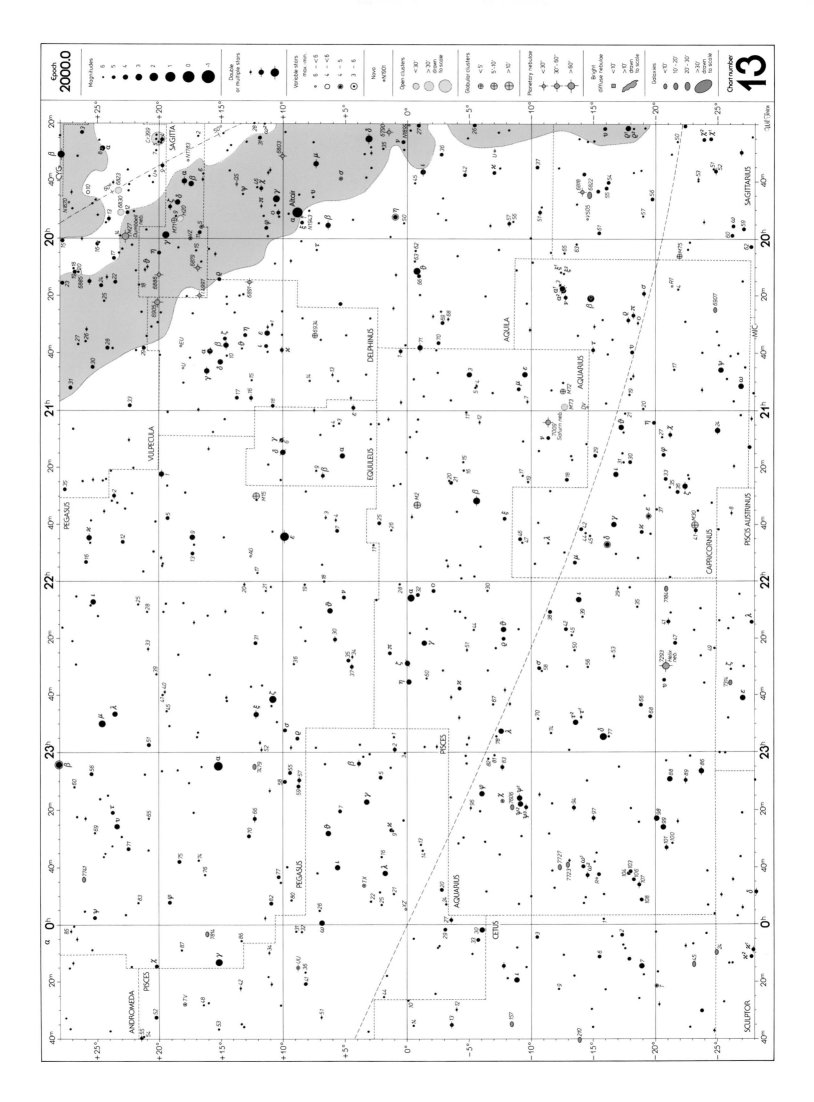

Chart 14   *RA 0h to 4h. Dec –20° to –70°*

Variable stars

		RA h	m	Dec °	Range	Type	Period d	Spectrum
R	Hor	02	53.9	–49 53	4.7–14.3	M	404.0	M
TW	Hor	03	12.6	–57 19	5.2– 5.9	SR	158	N
ζ	Phe	01	08.4	–55 15	3.9– 4.4	EA	1.67	B+B
R	Scl	01	27.0	–32 33	5.8– 7.7	SR	370	N
S	Scl	00	15.4	–32 03	5.5–13.6	M	365.3	M

Double stars

		RA h	m	Dec °	PA	Sep	Mag	
θ	Eri	02	58.3	–40 18	088	8.2	3.4, 4.5	
τ⁴	Eri	03	19.5	–21 45	AB 288 / AC 112	5.7 / 39.2	3.7, 9.2 / 10.5	
χ	Eri	01	56.0	–51 37	202	5.0	3.7,10.7	
ρ	Eri	01	39.8	–56 12	194	11.2	5.5, 5.8	Binary, 484 y
α	For	03	12.1	–28 59	298	4.0	4.0, 7.0	Binary, 314 y
γ¹	For	02	49.8	–24 34	AB 145 / AC 143	12.0 / 40.9	6.1,12.5 / 10.5	
η²	For	02	50.2	–35 51	014	5.0	5.9,10.1	
χ³	For	03	28.2	–35 51	248	6.3	6.5,10.5	
ϖ	For	02	33.8	–28 14	244	10.8	5.0, 7.7	
β	Phe	01	06.1	–46 43	346	1.4	4.0, 4.2	
η	Phe	00	41.8	–57 28	217	19.8	4.4,11.4	
ξ	Phe	00	41.8	–56 30	253	13.2	5.8,10.2	
ε	Scl	01	45.6	–25 03	028	4.7	5.4, 8.6	Binary, 1192 y
ζ	Scl	00	02.3	–29 43	320	3.0	5.0,13.0	
κ¹	Scl	00	09.3	–27 59	265	1.4	6.1, 6.2	
λ¹	Scl	00	42.7	–38 28	003	0.7	6.7, 7.0	
β¹	Tuc	00	31.5	–62 58	169	27.1	4.4, 4.8	
β²	Tuc	00	31.6	–62 58	295	0.6	4.8, 6.0	Binary, 44.4 y
κ	Tuc	01	15.8	–68 53	336	5.4	5.1, 7.3	

Globular clusters

NGC		RA h	m	Dec °	Diam	Mag
288	Scl	00	52.8	–26 35	13.8	8.1
1261	Hor	03	12.3	–55 13	6.9	8.4

Planetary nebula

NGC		RA h	m	Dec °	Diam	Mag	Mag*
1360	For	03	33.3	–25 51	390	—	11.3

Galaxies

NGC		RA		Dec	Mag	Diam	Type
24	Scl	00	09.9	–24 58	11.5	5.5× 1.6	Sb
45	Cet	00	14.1	–23 11	10.4	8.1× 5.8	S
55	Scl	00	14.9	–39 11	8.2	32.4× 6.5	SB
134	Scl	00	30.4	–33 15	10.1	8.1× 2.6	SBb
247	Cet	00	47.1	–20 46	8.9	20.0× 7.4	S
253	Scl	00	47.6	–25 17	7.1	25.1× 7.4	Scp
300	Scl	00	54.9	–37 41	8.7	20.0×14.8	Sd
578	Cet	01	30.5	–22 40	10.9	4.8× 3.2	Sc
613	Scl	01	34.3	–29 25	10.0	5.8× 4.6	SBb
685	Eri	01	47.8	–52 47	11.8	4.1× 4.0	SBc
908	Cet	02	23.1	–21 14	10.2	5.5× 2.8	Sc
986	For	02	33.6	–39 02	11.0	3.7× 2.8	SBb
1097	For	02	46.3	–30 17	9.2	9.3× 6.6	SBb
1187	Eri	03	02.6	–22 52	10.9	5.0× 4.1	SBc
1201	For	03	04.1	–26 04	10.6	4.4× 2.8	SO
1249	Hor	03	10.1	–53 21	11.7	5.2× 2.7	SBc
1255	For	03	13.5	–25 44	11.1	4.1× 2.8	Sa
1291	Eri	03	17.3	–41 08	8.5	10.5× 9.1	SBa
1302	For	03	19.9	–26 04	11.5	4.4× 4.2	SBa
1313	Ret	03	18.3	–66 30	9.4	8.5× 6.6	SBd
1316	For	03	22.7	–37 12	8.8	7.1× 5.5	SBOp
1326	For	03	23.9	–36 28	10.5	4.0× 3.0	SBO
1332	Eri	03	26.3	–21 20	10.3	4.6× 1.7	E7
1344	For	03	28.3	–31 04	10.3	3.9× 2.3	E3
1350	For	03	31.1	–33 38	10.5	4.3× 2.4	SBb
1365	For	03	33.6	–36 08	9.5	9.8× 5.5	SBb
1371	For	03	35.0	–24 56	11.5	5.4× 4.0	SBa
1380	For	03	36.5	–34 59	11.1	4.9× 1.9	SO
1385	For	03	37.5	–24 30	11.2	3.0× 2.0	Sc
1395	Eri	03	38.5	–23 02	11.3	3.2× 2.5	E3
1398	For	03	38.9	–26 20	9.7	6.6× 5.2	SBb
1399	For	03	38.5	–35 27	9.9	3.2× 3.1	E1p
1404	For	03	38.9	–35 35	10.2	2.5× 2.3	E1
1411	Hor	03	38.8	–44 05	11.9	2.8× 2.3	SO
1425	For	03	42.2	–29 54	11.7	5.4× 2.7	Sb
1433	Hor	03	42.0	–47 13	10.0	6.8× 6.0	SBa
1448	Hor	03	44.5	–44 39	11.3	8.1× 1.8	Sc
1493	Hor	03	57.5	–46 12	11.8	2.6× 2.3	SBc

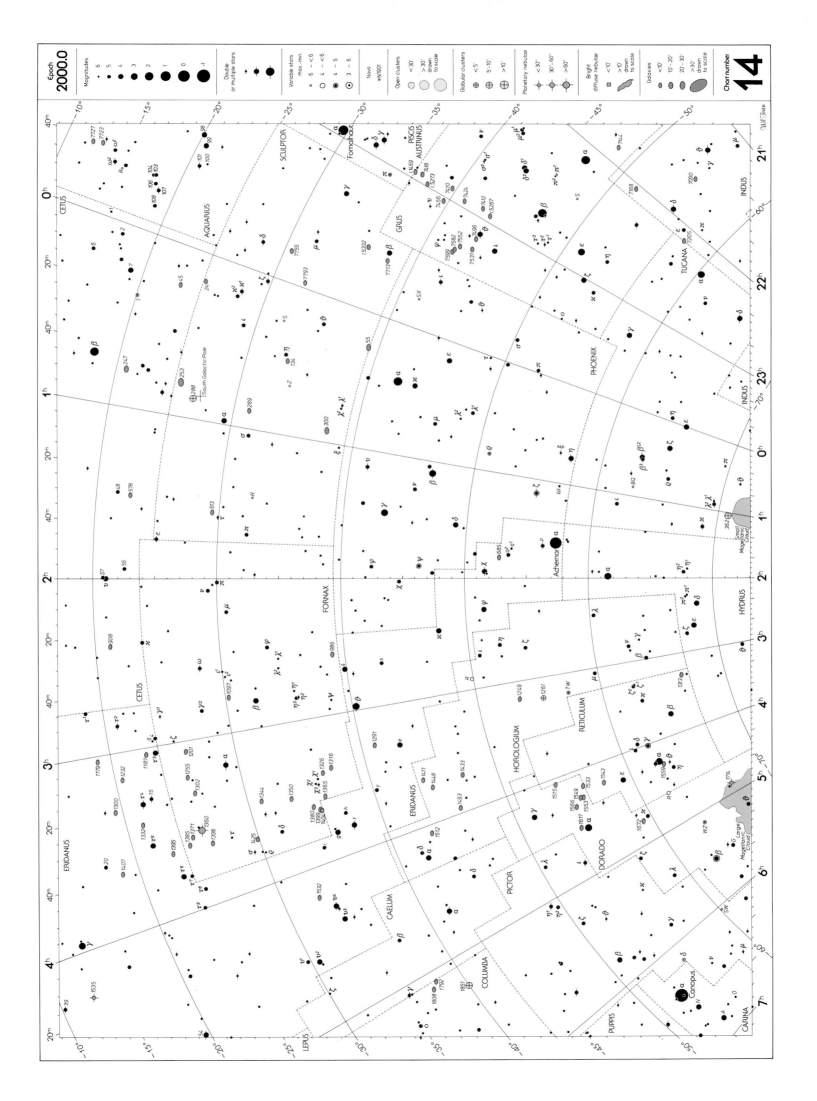

Chart 15 RA 4h to 8h. Dec −20° to −70°

Variable stars

		RA h m	Dec °	Range	Type	Period d	Spectrum
EW	CMa	07 14.3	−26 21	4.4–4.8	Irr	—	B
UW	CMa	07 18.7	−24 34	4.8–5.3	EB	4.39	07
β	Dor	07 33.6	−62 29	3.7–4.1	Cep	9.84	F-G
R	Dor	04 36.8	−62 05	4.8–6.6	SR	338	M
L³	Pup	07 13.5	−44 39	2.6–6.2	SR	140	M
V	Pup	07 58.2	−49 15	4.7–5.2	EB	1.45	B+B

Double stars

		RA	Dec	PA	Sep	Mag
α	Cae	04 40.6	−41 52	121	6.6	4.2,12.5
γ	Cae	05 04.4	−35 29	308	2.9	4.6, 8.1
ε	CMa	06 58.6	−28 58	161	7.5	1.5, 7.4
ξ¹	CMa	06 31.9	−23 25	144	24.6	4.3,13.9
π	CMa	06 55.6	−20 08	018	11.6	4.7, 9.7
17	CMa	06 55.0	−20 24	AB 147 / AC 184 / AD 186	44.4 / 50.5 / 129.9	5.8, 9.3 / 9.0 / 9.5
α	Col	05 39.6	−34 04	359	13.5	2.6,12.3
γ	Col	05 57.5	−35 17	110	33.8	4.4,12.7
π²	Col	06 07.9	−42 09	150	0.1	6.2, 6.3
α	Dor	04 34.0	−55 03	AB 182 / AB+C 101	0.2 / 77.7	3.8, 4.3 / 9.8
υ⁴	Eri	04 17.9	−33 48	AB+C 013	49.2	3.6,11.8 A is a close double
β	Lep	05 28.2	−20 46	AB 330 / AC 145 / AD 075 / AE 058	2.5 / 64.3 / 206.4 / 241.5	2.8, 7.3 / 11.8 / 10.3 / 10.3
η¹	Pic	05 02.8	−49 09	198	10.6	5.4,13.0
θ	Pic	05 24.8	−52 19	AB 152 / AB+C 287	0.2 / 38.2	6.9, 7.2 / 6.8
μ	Pic	06 32.0	−58 45	231	2.4	5.8, 9.0
σ	Pup	07 29.2	−43 18	074	22.3	3.3, 9.4
1	Pup	07 43.5	−28 24	033	26.2	4.6,13.5

Open clusters

NGC		RA	Dec	Diam	Mag	N*	
2287	CMa	06 46.0	−20 44	38	4.5	80	M41
2362	CMa	07 17.8	−24 57	8	4.1	60	τ CMa cluster
2383	Pup	07 24.8	−20 56	6	8.4	40	
2421	Pup	07 36.3	−20 37	10	8.3	70	
2439	Pup	07 40.8	−31 39	10	6.9	80	R Pup Asterism
2447	Pup	07 44.6	−23 52	22	6.2	80	M93
2451	Pup	07 45.4	−37 58	45	2.8	40	
2455	Pup	07 49.0	−21 18	8	10.2	50	
2477	Pup	07 52.3	−38 33	27	5.8	160	
2489	Pup	07 56.2	−30 04	8	7.9	45	
2516	Car	07 58.3	−60 52	30	3.8	80	

Globular clusters

NGC		RA	Dec	Diam	Mag	
1851	Col	05 14.1	−40 03	11.0	7.3	X-ray source
1904	Lep	05 24.5	−24 33	8.7	8.0	M79

Nebulae

NGC		RA	Dec	Diam	Mag*	
2070	Dor	05 38.7	−69 06	40×25	—	30 Doradus in LMC
2467	Pup	07 52.5	−26 24	8×7	9.2	Gum 9

Galaxies

NGC		RA	Dec	Mag	Diam	Type
1512	Hor	04 03.9	−43 21	10.6	4.0×3.2	SBa
1532	Eri	04 12.1	−32 52	11.1	5.6×1.8	Sb
1553	Dor	04 16.2	−55 47	9.5	4.1×2.8	SO
1559	Ret	04 17.6	−62 47	10.4	3.3×2.1	SBc
1549	Dor	04 15.7	−55 36	9.9	3.7×3.2	EO
1617	Dor	04 31.7	−54 36	10.4	4.7×2.4	SBa
1672	Dor	04 45.7	−59 15	11.0	4.8×3.9	SBb
1744	Lep	05 00.0	−26 01	11.2	6.8×4.1	SBc
1792	Col	05 05.2	−37 59	10.2	4.0×2.1	Sb
1808	Col	05 07.7	−37 32	9.9	7.2×4.1	SBa
1964	Lep	05 33.4	−21 57	10.8	6.2×2.5	Sb
2090	Col	05 47.0	−34 14	11.7	4.5×2.3	Sc
2207	CMa	06 16.4	−21 22	10.7	4.3×2.9	Sc
2217	CMa	06 21.7	−27 14	10.4	4.8×4.4	SBa
2223	CMa	06 24.6	−22 50	11.4	3.3×3.0	SBb
2280	CMa	06 44.8	−27 38	11.8	5.6×3.2	Sb
2442	Vol	07 36.4	−69 32	11.2	6.0×5.5	SBb
LMC	Dor	05 23.6	−69 45	0.1	650×550	SBm

Large Magellanic Cloud. Contains 30 Doradus and three planetary nebulae, 1714, 1722 and 1743.

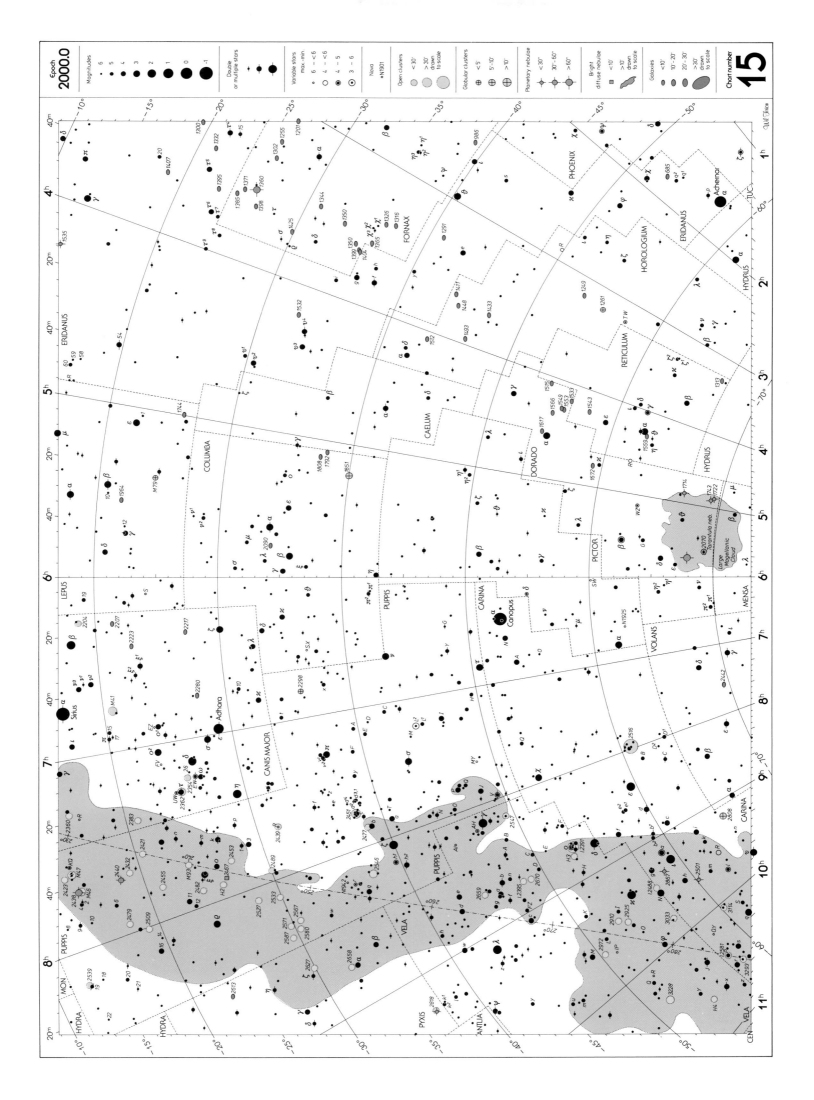

Chart 16 *RA 8h to 12h. Dec −20° to −70°*

Variable stars

		RA h m	Dec °	Range	Type	Period d	Spectrum
S	Ant	09 32.3	−28 38	6.4– 6.9	EW	0.65	A
U	Ant	10 35.2	−39 44	5.7– 6.8	Irr	—	N
η	Car	10 45.1	−59 41	−0.8– 7.9	Irr	—	Pec
R	Car	09 32.2	−62 47	3.9–10.5	M	308.7	M
U	Car	10 57.8	−59 44	5.7– 7.0	Cep	38.77	F-G
ZZ(l)	Car	09 45.2	−62 30	3.3– 4.2	Cep	35.53	F-K
S	Pyx	09 05.1	−23 05	8.0–14.2	M	206.4	M
Z	Vel	09 52.9	−54 11	7.8–14.8	M	421.6	M
AH	Vel	08 12.0	−46 29	5.5– 5.9	Cep	4.23	F
AI	Vel	08 14.1	−44 34	6.4– 7.1	δSct	0.11	A-F

Double stars

		RA	Dec	PA	Sep	Mag
δ	Ant	10 29.6	−30 36	226	11.0	5.6, 9.6
θ	Ant	09 44.2	−27 46	005	0.1	5.4, 5.6
π	Cen	11 21.0	−54 29	128	0.4	4.3, 5.0 Binary, 39.2 y
β	Hya	11 52.9	−33 54	008	0.9	4.7, 5.5
44	Hya	10 34.0	−23 45	061	19.1	5.1,13.8
δ	Pyx	08 55.5	−27 41	AB 268	23.8	4.9,14.0
				CD 017	2.5	11.0,11.0
ε	Pyx	09 09.9	−30 22	A+BC 147	17.8	5.6,10.5
				BC 088	0.3	10.5,10.8
				AD 340	35.4	5.6,13.5
ζ	Pyx	08 39.7	−29 34	061	52.4	4.9, 9.1
η	Pyx	08 37.9	−26 15	097	16.0	5.3,13.1
χ	Pyx	09 08.0	−25 52	263	2.1	4.6, 9.8
γ	Vel	08 09.5	−47 20	AB 220	41.2	1.9, 4.2
				AC 151	62.3	8.2
				AD 141	93.5	9.1
				DE 146	1.8	12.5
δ	Vel	08 44.7	−54 53	AB 153	2.6	2.1, 5.1
				AC 061	69.2	11.0
				CD 102	6.2	13.5
μ	Vel	10 46.8	−49 25	055	2.3	2.7, 6.4 Binary, 116 y
b	Vel	08 40.6	−46 39	058	37.5	3.8,10.2

Open clusters

NGC		RA	Dec	Diam	Mag	N*	
2516	Car	07 58.3	−60 52	30	3.8	80	
2527	Pup	08 05.3	−28 10	22	6.5	40	
2533	Pup	08 07.0	−29 54	3.5	7.6	60	
2546	Pup	08 12.4	−37 38	41	6.3	40	
2547	Vel	08 10.7	−49 16	20	4.7	80	
2567	Pup	08 18.6	−30 38	10	7.4	40	
2571	Pup	08 18.9	−29 44	13	7.0	30	
2580	Pup	08 21.6	−30 19	8	9.7	50	
2587	Pup	08 23.5	−29 30	9	9.2	40	
2627	Pyx	08 37.3	−29 57	11	8.4	60	
2658	Pyx	08 43.4	−32 39	12	9.2	80	
2669	Vel	08 44.9	−52 58	12	6.1	40	
2670	Vel	08 45.5	−48 47	9	7.8	30	
2910	Vel	09 30.4	−52 54	5	7.2	30	
2925	Vel	09 33.7	−53 26	12	8.3	40	
2972	Vel	09 40.3	−50 20	4	9.9	25	
3033	Vel	09 48.8	−56 25	5	8.8	50	
3114	Car	10 02.7	−60 07	35	4.2	35	
3228	Vel	10 21.8	−51 43	18	6.0	15	
3532	Car	11 06.4	−58 40	55	3.0	150	
3572	Car	11 10.4	−60 14	7	6.6	35	
3590	Car	11 12.9	−60 47	4	8.2	25	
3680	Car	11 25.7	−43 15	12	7.6	30	
3766	Cen	11 36.1	−61 37	12	5.3	100	
3960	Cen	11 50.9	−55 42	7	8.3	45	
IC 2391	Vel	08 40.2	−53 04	50	2.5	30	o Velorum cluster
IC 2395	Vel	08 41.1	−48 12	8	4.6	40	
IC 2488	Vel	09 27.6	−56 59	15	7.4	70	
IC 2581	Car	10 27.4	−57 38	8	4.3	25	
IC 2602	Car	10 43.2	−64 24	50	1.9	60	θ Carinae cluster
IC 2714	Car	11 17.9	−62 42	12	8.2	100	
IC 2944	Cen	11 36.6	−63 02	15	4.5	30	λ Centauri cluster
Mel 101	Car	10 42.1	−65 06	14	8.0	50	
Mel 105	Car	11 19.5	−63 30	4	8.5	70	

Globular clusters

NGC		RA	Dec	Diam	Mag	
2808	Car	09 12.0	−64 52	13.8	6.3	
3201	Vel	10 17.6	−46 25	18.2	6.7	Dun 445

Planetary nebulae

NGC		RA	Dec	Diam	Mag	Mag*
2818	Pyx	09 16.0	−36 28	38	13.0	13.0
2867	Car	09 21.4	−58 19	11	9.7	13.6
3132	Car	10 07.7	−40 26	47	8.2	10.1
3211	Car	10 17.8	−62 40	12	11.8	—
3918	Cen	11 50.3	−57 11	12	8.4	10.8
IC 2448	Car	09 07.1	−69 57	8	11.5	12.9
IC 2501	Car	09 38.8	−60 05	25	11.3	—
IC 2621	Car	11 00.3	−65 15	5	—	13.6

Galaxies

NGC		RA	Dec	Mag	Diam	Type
2613	Pyx	08 33.4	−22 58	10.4	7.2×2.1	Sb
2784	Hya	09 12.3	−24 10	10.1	5.1×2.3	S0
2835	Hya	09 17.9	−22 21	11.1	6.3×4.4	Sp
2997	Ant	09 45.6	−31 11	10.6	8.1×6.5	Sc
3109	Hya	10 03.1	−26 09	10.4	14.5×3.5	Irr
3223	Ant	10 21.6	−34 16	11.8	4.1×2.6	Sb
3347	Ant	10 42.8	−36 22	12.5	4.4×2.6	SBb
3511	Crt	11 03.4	−23 05	11.6	5.4×2.2	Sc
3513	Crt	11 03.8	−23 15	12.0	2.8×2.3	SBc
3557	Cen	11 10.0	−37 32	10.4	4.0×2.7	E3
3585	Hya	11 13.3	−26 45	10.0	2.9×1.6	E5
3621	Hya	11 18.3	−32 49	9.9	10.0×6.5	Sc
3923	Hya	11 51.0	−28 48	10.1	2.9×1.9	E3

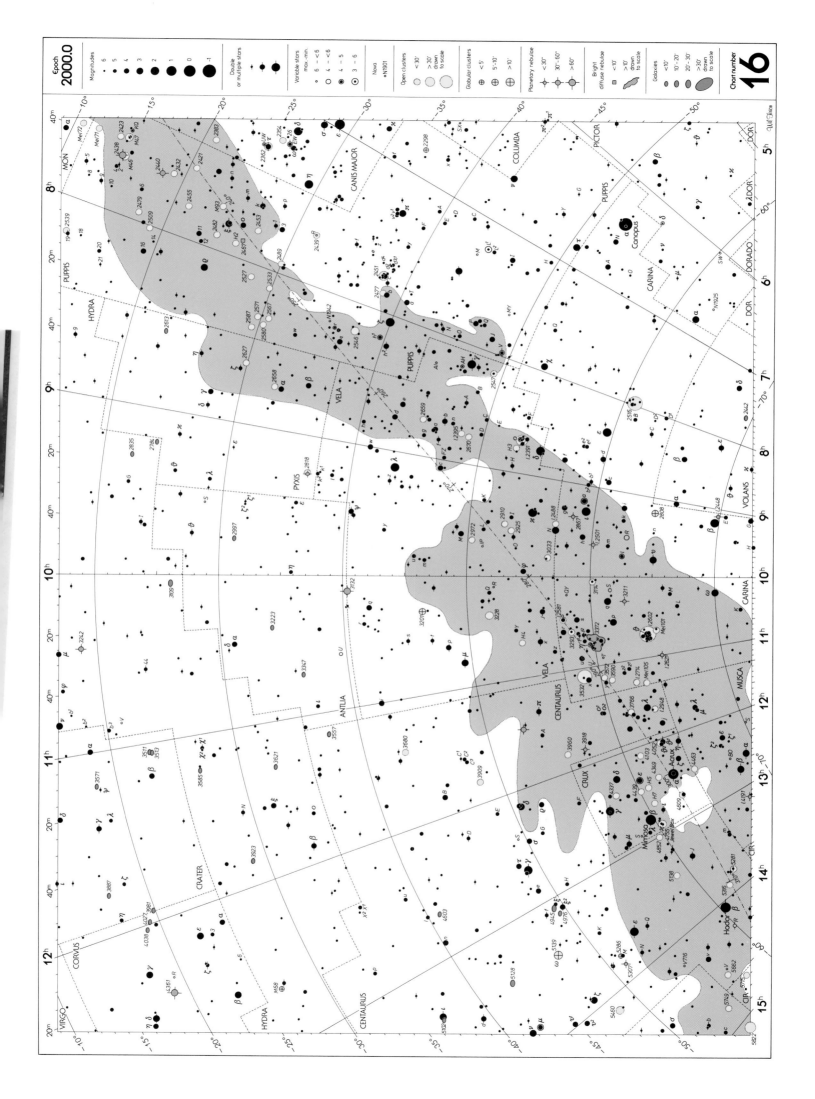

Chart 17   RA 12h to 16h. Dec −20° to −70°

Variable stars

		RA h m	Dec °	Range	Type*	Period d	Spectrum
μ	Cen	13 49.6	−42 28	2.9– 3.5	Irr	—	B
R	Cen	14 16.6	−59 55	5.3–11.8	M	546.2	M
S	Cen	12 24.6	−49 26	6.0– 7.0	SR	65	N
T	Cen	13 41.8	−33 36	5.5– 9.0	SR	90	K–M
V	Cen	14 32.5	−56 53	6.4– 7.2	Cep	5.49	F–G
W	Cen	11 55.0	−59 15	7.6–13.7	M	201.6	M
S	Cru	12 54.4	−58 26	6.2– 6.9	Cep	4.69	F–G
R	Hya	13 29.7	−23 17	4.0–10.0	M	389.6	M
GG	Lup	15 18.9	−40 47	5.4– 6.0	EB	2.16	B+A
R	Mus	12 42.1	−69 24	5.9– 6.7	Cep	7.48	F
R	Nor	15 36.0	−49 30	6.5–13.9	M	492.7	M
T	Nor	15 44.1	−54 59	6.2–13.6	M	242.6	M

Double stars

		RA h m	Dec °	PA	Sep	Mag	
α	Cen	14 39.6	−60 50	215	19.7	0.0, 1.2	Binary, 79.9 y
β	Cen	14 03.8	−60 22	251	1.3	0.7, 3.9	
γ	Cen	12 41.5	−48 58	353	1.0	2.9, 2.9	Binary, 84.5 y
ε	Cen	13 39.9	−53 28	158	36.0	2.3,12.7	
η	Cen	14 35.5	−42 09	270	5.0	2.6,13.5	
3	Cen	13 51.8	−33 00	108	7.9	4.5, 6.0	
4	Cen	13 53.2	−31 56	185	14.9	4.8, 8.4	
α	Cir	14 42.5	−64 59	232	15.7	3.2, 8.6	
γ	Cir	15 23.4	−59 19	033	0.6	5.1, 5.5	Binary, 180 y
δ	Cir	15 16.9	−60 57	270	50.0	5.1,13.4	
ζ	Crv	12 20.6	−22 13	066	11.2	5.2,13.6	
α	Cru	12 26.6	−63 06	AB 115	4.4	1.4, 1.9	
				AC 202	90.1	1.0, 4.9	
γ	Cru	12 31.2	−57 07	AB 031	110.6	1.6, 6.7	
				AC 082	155.2	9.5	
η	Cru	12 06.9	−64 37	299	44.0	4.2,11.7	
θ¹	Cru	12 03.0	−63 19	325	4.5	4.3,13.6	
ι	Cru	12 45.6	−60 59	022	26.9	4.7, 9.5	
μ	Cru	12 54.6	−57 11	017	34.9	4.0, 5.2	
R	Hya	13 29.7	−23 17	324	21.2	var,12.0	
52	Hya	14 28.2	−29 30	AB 130	0.1	5.8, 5.8	
				AB+C 279	4.2	10.0	
				AB+D 282	140.8	12.0	
54	Hya	14 46.0	−15 27	126	8.6	5.1, 7.1	
59	Hya	14 58.7	−27 39	335	0.8	6.3, 6.6	
κ	Lup	15 11.9	−48 44	144	26.8	3.9, 5.8	
μ	Lup	15 18.5	−47 53	AB 142	1.2	5.1, 5.2	
				AC 130	23.7	7.2	
ξ	Lup	15 56.9	−33 58	049	10.4	5.3, 5.8	
π	Lup	15 05.1	−47 03	073	1.4	4.6, 4.7	
τ¹	Lup	14 26.1	−45 13	204	148.2	4.6, 9.3	
υ	Lup	15 24.7	−39 43	038	1.4	5.4,10.9	
α	Mus	12 37.2	−69 08	316	29.6	2.7,12.8	
β	Mus	12 46.3	−68 06	014	1.4	3.7, 4.0	
ζ²	Mus	12 22.1	−67 31	130	32.4	5.2,10.6	
θ	Mus	13 08.1	−65 18	187	5.3	5.7, 7.3	
π	Sco	15 58.9	−26 07	132	50.4	2.9,12.1	
2	Sco	15 53.6	−25 20	274	2.5	4.7, 7.4	

Open clusters

NGC		RA	Dec	Diam	Mag	N*
4103	Cru	12 06.7	−61 15	7	7.4	45
4337	Cru	12 23.9	−58 08	3.5	8.9	—
4439	Cru	12 28.4	−60 06	4	8.4	—
4349	Cru	12 24.5	−61 54	16	7.4	30
4463	Mus	12 30.0	−64 48	5	7.2	30
4609	Cru	12 42.3	−62 58	5	6.9	40
4755	Cru	12 53.6	−60 20	10	4.2	50+
5138	Cen	13 27.3	−59 01	8	7.6	40
5281	Cen	13 46.6	−62 54	5	5.9	40
5316	Cen	13 53.9	−61 52	14	6.0	80
5460	Cen	14 07.6	−48 19	25	5.6	40
5617	Cen	14 29.8	−60 43	10	6.3	80
5662	Cen	14 35.2	−56 33	12	5.5	70
5749	Lup	14 48.9	−54 31	8	8.8	30
5822	Lup	15 05.2	−54 21	40	6.5	150
5823	Cir	15 05.7	−55 36	10	7.9	100
5925	Nor	15 27.7	−54 31	15	8.4	120
5999	Nor	15 52.2	−56 28	5	9.0	40
H.5	Cru	12 29.0	−60 46	6	7.1	—

Globular clusters

NGC		RA	Dec	Diam	Mag	
4590	Hya	12 39.5	−26 45	12.0	8.2	M68
5139	Cen	13 26.8	−47 29	36.3	3.6	ω Centauri
5286	Cen	13 46.4	−51 22	9.1	7.6	
5824	Lup	15 04.0	−33 04	6.2	9.0	
5897	Lib	15 17.4	−21 01	12.6	8.6	
5927	Lup	15 28.0	−50 40	12.0	8.3	H IV 19
5986	Lup	15 46.1	−37 47	9.8	7.1	Dun 552

Planetary nebulae

NGC		RA	Dec	Diam	Mag	Mag*
5882	Lup	15 16.8	−45 39	7	10.5	12.0
IC 4191	Mus	13 08.8	−67 39	5	12.0	
IC 4406	Lup	14 22.4	−44 09	28	10.6	14.7

Nebulae

NGC		RA	Dec	Diam	Mag*
5367	Cen	13 57.7	−39 59	4×3	Includes IC 4347. Double nucleus
—	Cru	12 53	−63	400×300	Coal Sack: dark nebula, area 26.2 sq deg

Galaxies

NGC		RA	Dec	Mag	Diam	Type	
4603	Cen	12 40.9	−40 59	12.0	3.8 × 2.5	Sc	
4945	Cen	13 05.4	−49 28	9.5	20.0 × 4.4	SBc	
4976	Cen	13 08.6	−49 30	10.2	4.3 × 2.6	E4p	
5078	Hya	13 19.8	−27 24	12.0	3.2 × 1.7	Sa	
5101	Hya	13 21.8	−27 26	11.7	5.5 × 4.9	SBa	
5061	Hya	13 18.1	−26 50	11.7	2.6 × 2.3	E2	
5068	Vir	13 18.9	−21 02	10.8	6.9 × 6.3	SBc	
5085	Hya	13 20.3	−24 26	11.9	3.4 × 3.0	Sb	
5102	Cen	13 22.0	−36 38	9.6	9.3 × 3.5	S0	
5128	Cen	13 25.5	−43 01	7.0	18.2×14.3	S0p	Centaurus A
5236	Hya	13 37.0	−29 52	8.2	11.2×10.2	Sc	M83
5253	Cen	13 39.9	−31 39	10.6	4.0 × 1.7	E5	
5483	Cen	14 10.4	−43 19	12.0	3.1 × 2.8	Sc	
5643	Lup	14 32.7	−44 10	10.7	4.6 × 4.1	SB0	
IC 4296	Cen	13 36.6	−33 58	10.6	—	E0	

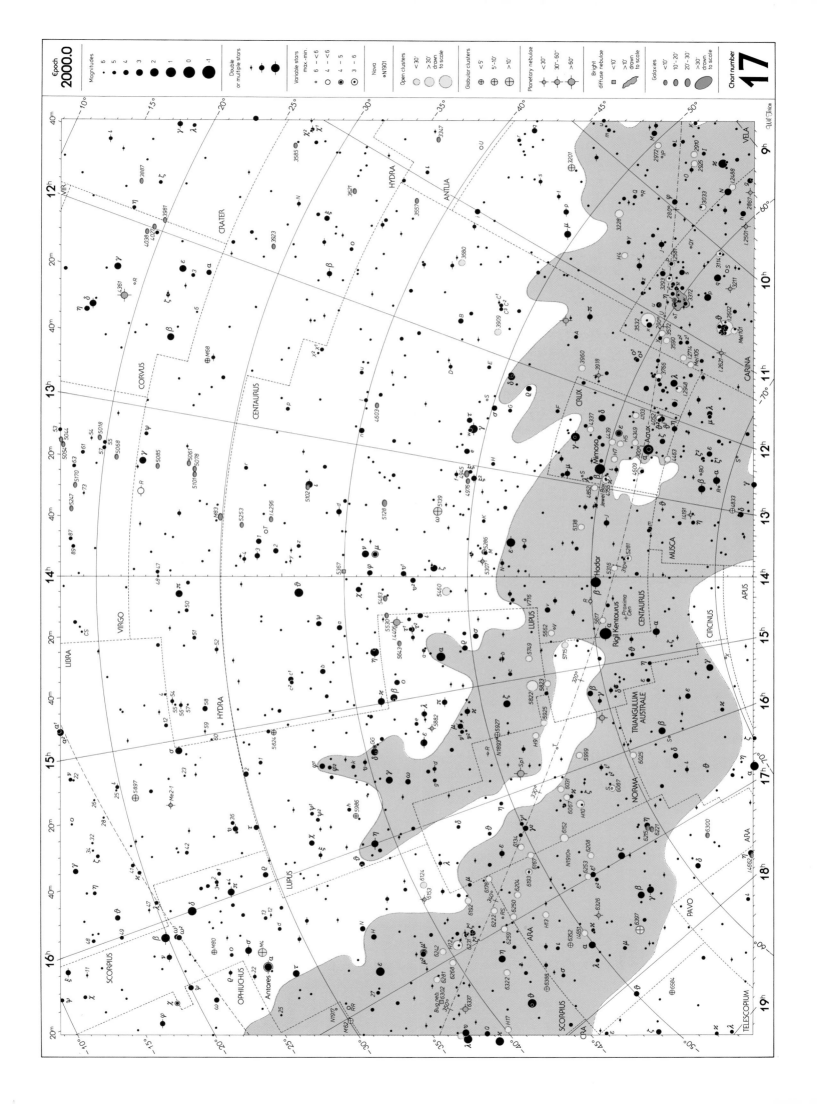

Chart 18 *RA 16h to 20h. Dec –20° to –70°*

Variable stars

		RA (h m)	Dec (°)	Range	Type	Period (d)	Spectrum	
ε	CrA	18 58.7	–37 06	4.7– 5.0	EW	0.59	F	
S	Nor	16 18.9	–57 54	6.1– 6.8	Cep	9.75	F-G	
κ	Pav	18 56.9	–67 14	3.9– 4.7	Cep.W	9.09	F	
λ	Pav	18 52.2	–62 11	3.4– 4.3	Irr	—	B	
S	Pav	19 55.2	–59 12	6.6–10.4	SR	386.0	M	
BM	Sco	17 41.0	–32 13	6.8– 8.7	SR	850.0	K	
RR	Sco	16 55.6	–30 35	5.0–12.4	M	279.4	M	
RS	Sco	16 56.6	–45 06	6.2–13.0	M	320.0	M	
RU	Sco	17 42.4	–43 45	7.8–13.7	M	369.2	M	
RR	Sgr	19 55.9	–29 11	5.6–14.0	M	334.6	M	
RU	Sgr	19 58.7	–41 51	6.0–13.8	M	240.3	M	
RY	Sgr	19 16.5	–33 31	6.0–15.0	RCrB	—	Gp	
W	Sgr	18 05.0	–29 35	4.3– 5.1	Cep	7.59	F-G	
S	TrA	16 01.2	–63 47	6.1– 6.8	Cep	6.32	F	

Double stars

		RA	Dec (°)	PA	Sep	Mag	
γ	Ara	17 25.4	–56 23	{ AB 328 / AC 066 }	17.9 / 41.6	3.3, 3, 10.3 / 4.8, 5.1	
γ	CrA	19 06.4	–37 04	109	1.3	4.8, 5.1 Binary, 120.4 y	
λ	CrA	18 43.8	–38 19	214	29.2	5.1, 9.7	
ε	Nor	16 27.0	–47 03	335	22.8	4.8, 7.5	
ι¹	Nor	18 03.5	–57 47	100	0.2	5.3, 5.5 Binary, 26.9 y	
36	Oph	17 15.3	–26 36	150	4.7	5.1, 5.1 Binary, 549 y	
ξ	Pav	18 23.2	–61 30	154	3.3	4.4, 8.6	
α	Sco	16 29.4	–26 26	273	2.7	1.2, 5.4 Antares, Binary, 878 y	
σ	Sco	16 21.2	–25 36	273	20.0	2.9, 8.5	
12	Sco	16 12.3	–28 25	073	4.0	5.9, 7.9	
β¹	Sgr	18 22.6	–44 28	077	28.3	3.9, 8.0 Wide naked-eye pair with β² (4.3 y)	
σ	Sgr	19 02.6	–29 53	320	0.3	3.3, 3.4	
η	Sgr	18 17.6	–36 46	105	3.6	3.2, 7.8	
π	Sgr	19 09.8	–21 01	{ AB 150 / AB+C 122 }	0.1 / 0.4	3.7, 3.7 / 5.9	
21	Sgr	18 25.3	–20 32	289	1.8	4.9, 7.4	
52	Sgr	19 36.7	–24 53	170	2.5	4.7, 9.2	

Open clusters

NGC		RA	Dec	Diam	Mag	N*	
5050	TrA	16 03.7	–60 30	12	5.1	60	
6067	Nor	16 13.2	–54 13	13	5.6	100	
6087	Nor	16 18.9	–57 54	12	5.4	40 S	
6124	Sco	16 25.6	–40 40	29	5.8	100	
6134	Nor	16 27.7	–49 09	7	7.2	—	
6167	Nor	16 34.4	–49 36	8	6.7	—	
6178	Sco	16 35.7	–45 38	4	7.2	12	
6193	Ara	16 41.3	–48 46	15	5.2	—	
6204	Ara	16 46.5	–47 01	5	8.2	45	
6208	Ara	16 49.5	–53 49	16	7.2	60	
6231	Sco	16 54.0	–41 48	15	2.6	—	Normac cluster
6242	Sco	16 55.6	–39 30	9	6.4	—	
6250	Ara	16 58.0	–45 48	8	5.9	60	
6281	Sco	17 04.8	–37 54	8	5.4	—	Nebulosity
6383	Sco	17 34.8	–32 34	5	5.5	40	
6405	Sco	17 40.1	–32 13	15	4.2	80	M6 Butterfly cluster
6416	Sco	17 44.4	–32 21	18	5.7	40	
6520	Sco	18 03.4	–27 54	6	7.6	60	In M20 (Trifid)
6530	Sgr	18 04.8	–24 20	15	4.6	—	In M8 (Lagoon)
6531	Sgr	18 04.6	–22 30	13	5.9	70	M21
6546	Sgr	18 07.2	–23 20	13	8.0	150	
6475	Sco	17 53.9	–34 49	80	3.3	80	M7
IC 4651	Ara	17 24.7	–49 57	12	6.9	30	
H.10	Nor	16 19.9	–54 59	30	—	15	
H.13	Ara	17 05.4	–48 11	15	—	—	

Globular clusters

NGC		RA	Dec	Diam	Mag	
6093	Sco	16 17.0	–22 59	8.9	7.2	M80
6121	Nor	16 23.6	–26 32	26.3	5.9	M4
6266	Oph	17 01.2	–30 07	14.1	6.6	M62
6273	Oph	17 02.6	–26 16	13.5	7.1	M19
6304	Oph	17 14.5	–29 28	6.8	8.4	
6352	Ara	17 25.5	–48 25	7.1	8.1	
6355	Oph	17 24.0	–26 21	5.0	9.6	
6362	Ara	17 31.9	–67 03	10.7	8.3	
6388	Sco	17 36.3	–44 44	8.7	6.8	
6397	Ara	17 40.7	–53 40	25.7	5.6	Dun 473
6541	CrA	18 08.0	–43 42	13.1	6.6	
6544	Sgr	18 07.3	–25 00	8.9	8.2	
6553	Sgr	18 09.3	–25 54	8.1	8.2	
6558	Sgr	18 10.3	–31 46	3.7	—	
6569	Sgr	18 13.6	–31 50	5.8	8.7	
6624	Sgr	18 23.7	–30 22	5.9	8.3	H150
6626	Sgr	18 24.5	–24 52	11.2	6.9	M28
6637	Sgr	18 31.4	–32 21	7.1	7.7	M69
6638	Sgr	18 30.9	–25 30	5.0	9.2	H151
6652	Sgr	18 35.8	–32 59	3.5	8.9	
6656	Sgr	18 36.4	–23 54	24.0	5.1	M22
6715	Sgr	18 55.1	–30 09	9.1	7.7	M54
6752	Pav	19 10.9	–59 59	20.4	5.4	
6681	Sgr	18 43.2	–32 18	7.8	8.1	M70
6809	Sgr	19 40.0	–30 58	19.0	6.9	M55

Planetary nebulae

NGC		RA	Dec	Diam	Mag	Mag*	
6153	Sco	16 31.5	–40 15	25	11.5	11.5	
6302	Sco	17 13.7	–37 06	50	12.8	12.8	Bug Nebula
6337	Sco	17 22.3	–38 29	48	—	—	
6629	Sgr	18 25.7	–23 12	15	12.2	14.7	
6644	Sgr	18 32.6	–25 08	3	—	12.8	
IC 1297	CrA	19 17.4	–39 37	7	11.9	15.9	RU CrA
IC 4699	Tel	18 18.5	–45 59	10	—	12.9v	

Nebulae

NGC		RA	Dec	Diam	Mag*	
6514	Sgr	18 02.6	–23 02	29×27	7.6	Trifid Nebula, M20
6523	Sgr	18 08.3	–24 23	90×40	—	Lagoon Nebula, M8

Galaxies

NGC		RA	Dec	Mag	Diam	Type
6215	Ara	16 51.1	–58 59	11.8	2.0 × 1.6	Sc
6221	Ara	16 52.8	–59 13	11.5	3.2 × 2.3	SBc
6684	Pav	18 49.0	–65 11	10.4	3.7 × 2.7	SB0
6744	Pav	19 09.8	–63 51	9.0	15.5 × 10.2	SBb
6753	Pav	19 11.4	–57 03	11.9	2.5 × 2.2	Sb
IC 4662	Pav	17 47.1	–64 38	11.4	2.2 × 1.4	Irr

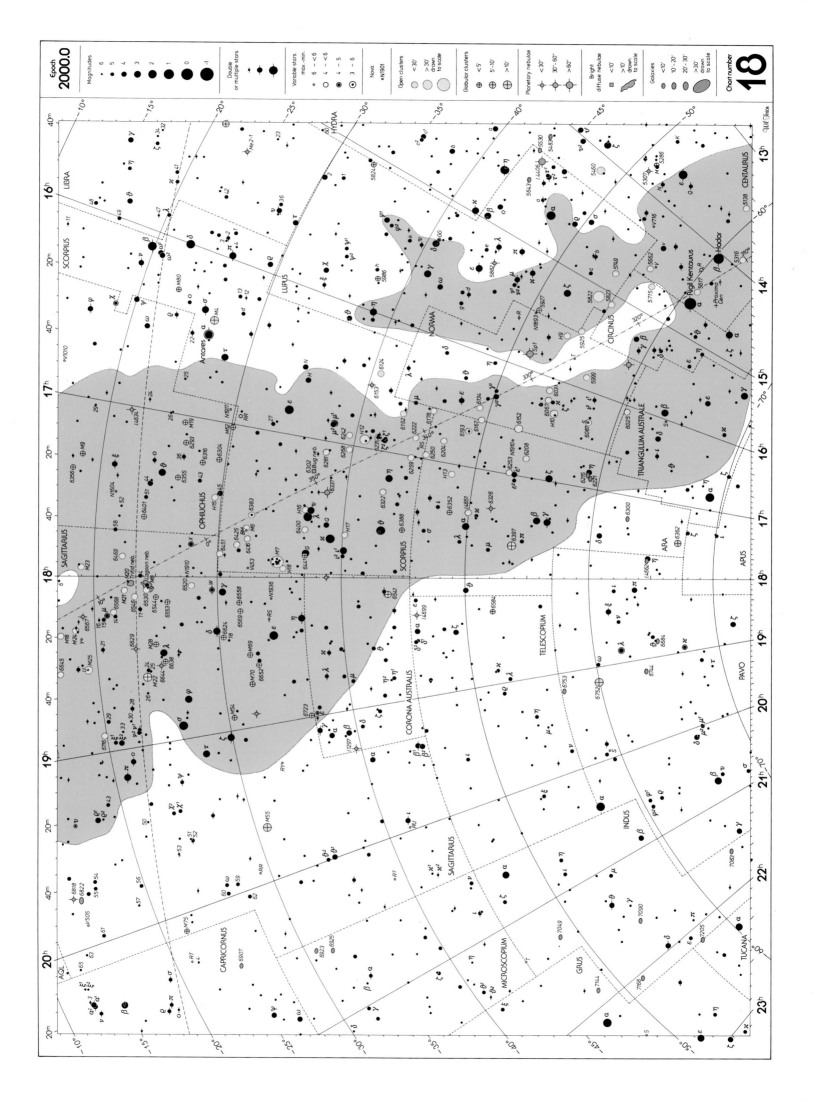

Chart 19 RA 20h to 6h. Dec −20° to −70°

Variable stars

		RA h m	Dec °	Range	Type	Period d	Spectrum	
RT	Cap	20 17.1	−21 19	8.9–11.7 pho.	SR	393	M	
S	Gru	22 26.1	−48 26	6.0–15.0	M	401.4	M	
T	Gru	22 25.7	−37 34	7.8–12.3	M	136.5	M	
T	Mic	20 27.9	−28 16	7.7– 9.6	SR	344	M	
Y	Pav	21 24.3	−69 44	5.7– 8.5	SR	233	N	
SX	Pav	21 28.7	−69 30	5.4– 6.0	SR	50	M	
RT	Sgr	20 17.7	−39 07	6.0–14.1	M	305.3	M	

Double stars

		RA	Dec	PA	Sep	Mag
41	Aqr	22 14.3	−21 04	AB 114 / AC 043	5.0 / 212.1	5.6, 7.1 / 9.0
86	Aqr	23 06.7	−23 45	083	2.9	4.5,14.5
89	Aqr	23 09.9	−22 27	007	0.4	5.1, 5.9
ζ	Cap	21 26.7	−22 25	013	21.3	3.7,12.3
24	Cap	21 07.1	−25 00	186	26.2	4.6,11.7
δ	Ind	21 57.9	−55 00	323	0.1	5.3, 5.3 Binary, 12.0 y
θ	Ind	21 19.9	−53 27	275	6.0	4.5, 7.0
θ	Gru	23 06.9	−43 31	075	1.1	4.5, 7.0
ν	Gru	23 06.9	−38 54	211	1.1	5.7, 8.0
α	Mic	20 50.0	−33 47	166	20.5	5.0,10.0
θ²	Mic	21 24.4	−41 00	AB 267 / AC 066	0.5 / 78.4	6.4, 7.0 / 10.5
β	PsA	23 31.5	−32 21	172	30.4	4.4, 7.9 Optical
γ	PsA	22 52.5	−32 53	262	4.2	4.5, 8.0
δ	PsA	22 55.9	−32 32	244	5.0	4.2, 9.2
η	PsA	22 00.8	−28 27	115	1.7	5.8, 6.8
6	PsA	21 32.2	−33 57	059	6.8	6.0,13.3
8	PsA	21 36.2	−26 10	008	18.4	5.7,13.9
δ	Scl	23 48.9	−28 08	AB 243 / AC 297	3.3 / 74.3	4.5,11.5 / 9.3
κ²	Sgr	20 23.9	−42 25	234	0.8	6.0, 6.9
δ	Tuc	22 27.3	−64 58	282	6.9	4.5, 9.0

Globular clusters

NGC		RA	Dec	Diam	Mag	
6864	Sgr	20 06.1	−21 55	6	8.6	M75
7099	Cap	21 40.4	−23 11	11	7.5	M30

Planetary nebula

NGC		RA	Dec	Diam	Mag	Mag*	
7293	Aqr	22 29.6	−20 48	770	6.5	13.5	Helix Nebula

Galaxies

NGC		RA	Dec	Mag	Diam	Type	
6907	Cap	20 25.1	−24 49	11.3	3.4×3.0	SBb	
6923	Mic	20 31.7	−30 50	12.1	2.5×1.4	Sb	
6925	Mic	20 34.3	−31 59	11.3	4.1×1.6	Sb	
7049	Ind	21 19.0	−48 34	10.7	2.8×2.2	S0	
7083	Ind	21 35.7	−63 54	11.8	4.5×2.9	Sb	
7090	Ind	21 36.5	−54 33	11.1	7.1×1.4	SBc	
7144	Gru	21 52.7	−48 15	10.7	2.5×2.3	E0	
7168	Ind	22 02.1	−51 45	12.6	2.0×1.6	E3	
7172	PsA	22 02.0	−31 52	11.9	2.2×1.3	S	
7174	PsA	22 02.1	−31 59	12.6	1.3×0.7	S	
7184	Aqr	22 02.7	−20 49	12.0	5.8×1.8	Sb	
7205	Ind	22 08.5	−57 25	11.4	5.3×2.2	Sb	
7314	PsA	22 35.8	−26 03	10.9	4.6×2.3	Sc	Arp 14
7410	Gru	22 55.0	−39 40	10.4	5.5×2.0	SBa	
7412	Gru	22 55.8	−42 39	11.4	4.0×3.1	SBb	
7418	Gru	22 56.6	−37 02	11.4	3.3×2.8	SBc	
7424	Gru	22 57.3	−41 04	11.0	7.6×6.8	SBc	
7456	Gru	23 02.1	−39 35	11.9	5.9×1.8	Sc	
7496	Gru	23 09.8	−43 26	11.1	3.5×2.8	SBb	
7531	Gru	23 14.8	−43 36	11.3	3.5×1.5	Sb	
7552	Gru	23 16.2	−42 35	10.7	3.5×2.5	SBb	
7582	Gru	23 18.4	−42 22	10.6	4.6×2.2	SBb	
7599	Gru	23 19.3	−42 15	11.4	4.4×1.5	Sc	
7713	Scl	23 36.5	−37 56	11.6	4.3×2.0	SBd	
7755	Scl	23 47.9	−30 31	11.8	3.7×3.0	SBb	
7793	Scl	23 57.8	−32 35	9.1	9.1×6.6	Sd	
IC 1459	Gru	22 57.2	−36 28	10.0	—	E3	
IC 5267	Scl	22 57.2	−43 24	10.5	5.0×4.1	S0	
IC 5273	Gru	22 59.5	−37 42	11.4	2.9×2.1	SBc	
IC 5332	Scl	23 34.5	−36 06	10.6	6.6×5.1	Sd	

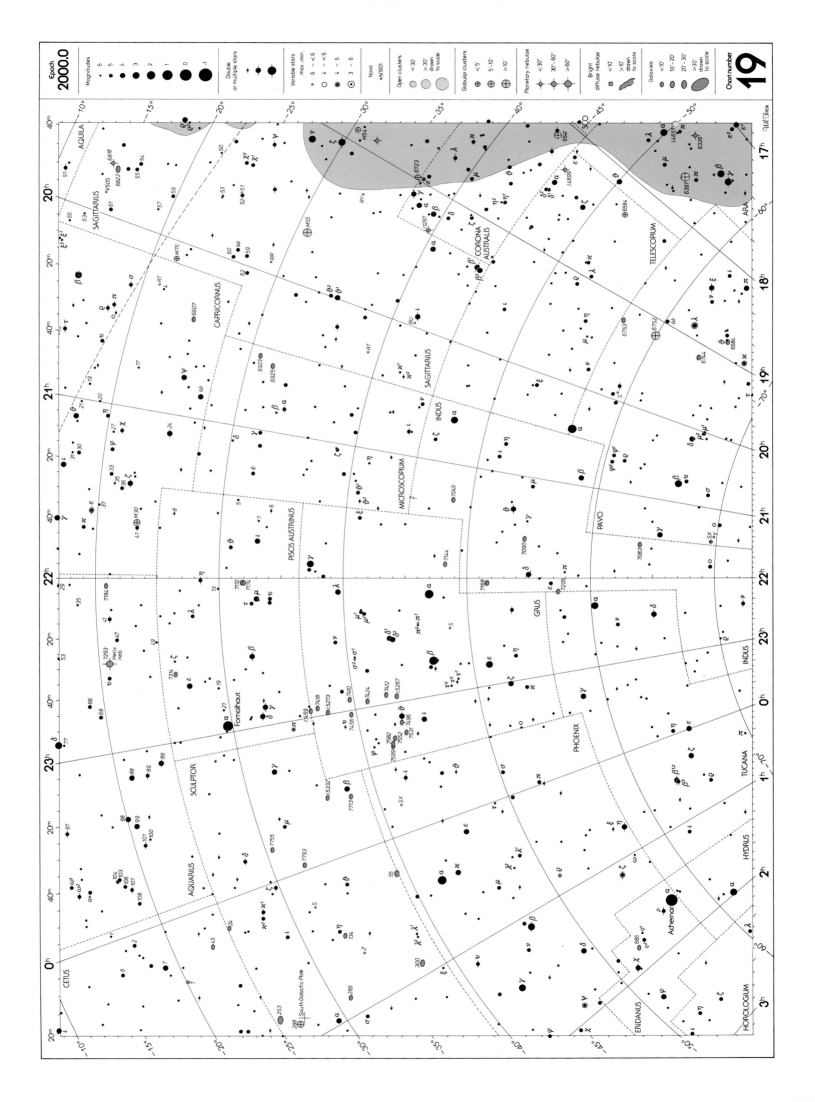

Chart 20 *Far south; declination below –70°*

Variable stars

		RA h m	Dec °	Range	Type	Period d	Spectrum
θ	**Aps**	14 05.3	–76 48	6.4– 8.6	**SR**	119	M
RS	**Cha**	08 43.2	–79 04	6.0– 6.7	**EA+δ Sct**	1.67	A-F
TZ	**Men**	05 30.2	–84 47	6.2– 6.9	**EA**	8.57	B
ε	**Oct**	22 00.0	–80 26	4.9– 5.4	**SR**	55	M
R	**Oct**	05 26.1	–86 23	6.4–13.2	**M**	405.6	M

Double stars

		RA	Dec	PA	Sep	Mag
δ	**Aps**	16 20.3	–78 42	012	102.9	4.7, 5.1
δ¹	**Cha**	10 45.3	–80 28	076	0.6	6.1, 6.4
ε	**Cha**	11 59.6	–78 13	188	0.9	5.4, 6.0
θ	**Cha**	08 20.6	–77 29	250	31.0	4.4,12.1
τ²	**Hyi**	01 47.8	–80 11	028	39.7	6.1,13.5
γ	**Men**	05 31.9	–76 20	107	38.2	5.2,11.2
ι	**Oct**	12 55.0	–85 07	230	0.6	6.0, 6.5
λ	**Oct**	21 50.9	–82 43	070	3.1	5.4, 7.7
μ²	**Oct**	20 41.7	–75 21	017	17.4	7.1, 7.6
γ²	**Vol**	07 08.8	–70 30	300	13.6	4.0, 5.9
ζ	**Vol**	07 41.8	–72 36	116	16.7	4.0, 9.8
θ	**Vol**	08 39.1	–70 23	AC 108	45.0	5.3,10.3 A is a close double
κ	**Vol**	08 19.8	–71 31	{ AB 057 { BC 030	65.0 37.7	5.4, 5.7 8.7

Globular clusters

NGC		RA	Dec	Diam	Mag	
104	**Tuc**	00 24.1	–72 05	30.9	4.0	47 Tucanae
362	**Tuc**	01 03.2	–70 51	12.9	6.6	
4372	**Mus**	12 25.8	–72 40	18.6	7.8	
4833	**Mus**	12 59.6	–70 53	13.5	7.3	
6101	**Aps**	16 25.8	–72 12	10.7	9.3	
IC 4499	**Aps**	15 00.3	–82 13	7.6	10.6	

Galaxies

NGC		RA	Dec	Mag	Diam	Type
3059	**Car**	09 50.2	–73 55	11.8	3.2×3.0	**SBb**
5967	**Aps**	15 48.1	–75 40	12.5	2.9×1.8	**SBc**
SMC	**TUC**	00 52.7	–72 50	2.3	280×160	Small cloud of Magellan

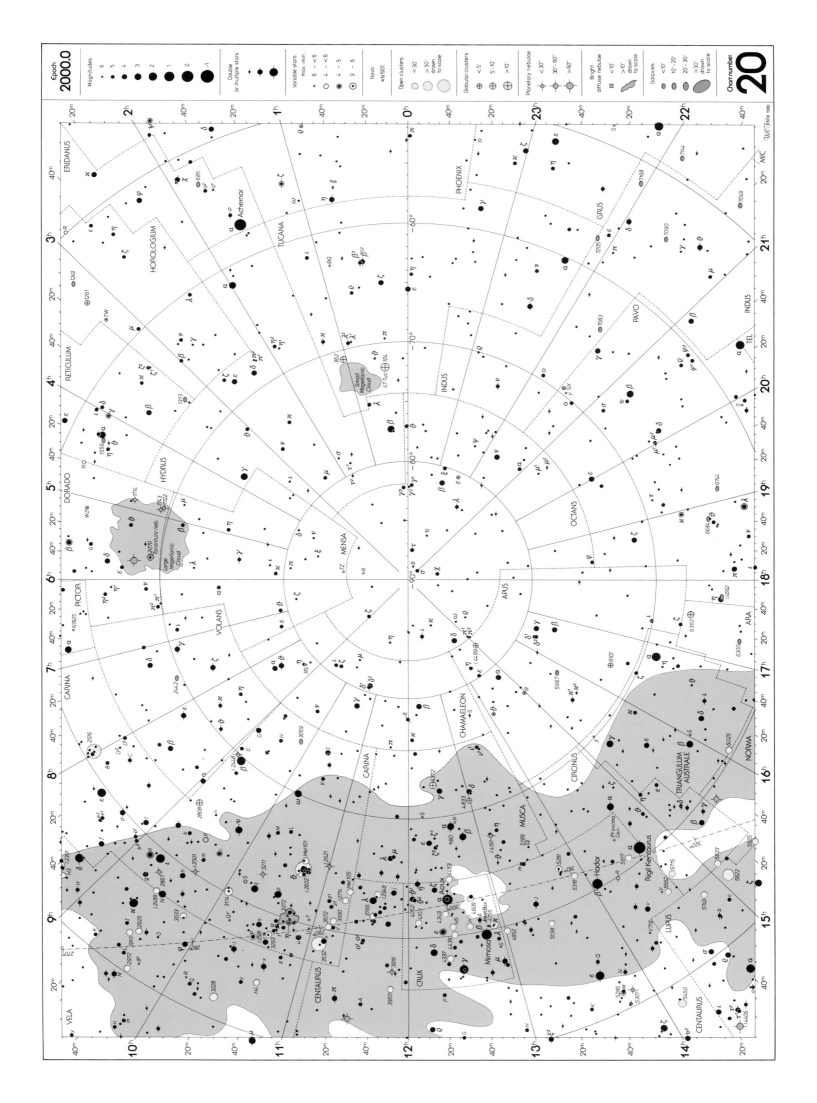

THE ALL-SKY MAPS

This final part of the *Cambridge Star Atlas* consists of six all-sky maps to show the general distribution of different objects in the sky. Each of the six maps shows the whole sky in a so-called 'equal-area' projection; the Mollweide projection, named after the man who invented it. 'Equal-area' means that in spite of the inevitable distortion the actual area covered by one square degree remains the same, no matter where on the map it is measured. So the distribution, or density of objects, is not influenced by the map's projection.

Unlike the main star charts, these maps show the galactic coordinates. The Galactic Equator is the central, horizontal line, marked 0°. On the main star charts it is represented as a line of dots and dashes, making an angle of almost 63° with the Celestial Equator. The Galactic Center (galactic longitude 0°, see atlas chart 18) lies close to the point where the constellations Sagittarius, Ophiuchus and Scorpius meet, and is in the center of the galactic maps in this chapter. At the top and bottom of the maps the North Galactic Pole (NGP) and the South Galactic Pole (SGP) are marked. They can be found on the star charts number 5 (in Coma Berenices) and 14 (in Sculptor) respectively.

The constellations

The first map gives the positions of the constellations, against this unusual grid. Stars down to magnitude 4.5 are plotted, plus some fainter stars to complete the constellation patterns, as on the monthly sky maps. In order not to overload the map, only the three-letter abbreviations of the constellation names are used. You can find the full names in Table C, in the introduction to the star charts.

Distribution of open clusters

The second map shows the open star clusters plotted as yellow disks against the background of stars and constellations (in grey and light blue). As explained

in the introduction to the main star charts the open clusters are found near the plane of the Milky Way, so on the map you will find most of these objects close to the Galactic Equator. The map shows all clusters that are plotted on the star charts. This is also the case with the globular clusters, the planetary nebulae and the galaxies.

Distribution of globular clusters

Map number three shows the globular star clusters, using the same yellow symbols that are used on the star charts. You can see that the distribution is very different from the open clusters. The globulars form a huge halo around the Milky Way. So they are much more scattered than the open clusters. But since we are not in the center of our galaxy (in fact, we are closer to the edge than to the center) we see most globular clusters in the direction of the galactic heart (in the center of the map). In the opposite direction (near 180° galactic longitude, the left and right hand edges of the map) they are almost absent.

Distribution of planetary nebulae

The distribution of planetary nebulae (drawn as green disks with four spikes) does not show any similarity with that of the open or the globular clusters. They are not exclusively found in the spiral arms of the Milky Way, like the open star clusters, nor in a halo, like the globular clusters. They fill a disk-like area, that is much thicker than the main Milky Way disk, in which the open clusters and the diffuse nebulae are found.

Distribution of diffuse nebulae

Because on the star charts of the *Cambridge Star Atlas* only a limited number of bright diffuse nebulae are drawn, a number of additional nebulae have been added, to show their distribution. The dark green squares are the ones drawn in the *Cambridge Star Atlas,* while the lighter green squares represent the

additional nebulae, taken from *Sky Atlas 2000.0*. As
you can see, the distribution of the nebulae is quite
similar to that of the open star clusters. It brings us
back to the disk of the galaxy. Like the open clusters,
the diffuse nebulae are found along the spiral arms of
the Milky Way. In some areas you see gaps, where
the nebulae are almost absent, but this is only an
illusion. Large areas of dark nebulae and clouds of
dust block the light of many remote bright nebulae
as well as that of other objects.

Distribution of galaxies

Since the galaxies, shown as red ovals, are not part
of the Milky Way, but are in fact very distant 'Milky
Ways' themselves, there is no relation between their
distribution in the sky and the distribution of objects
that belong to our own Milky Way. But, looking at
the map, you will get the impression that there is a
relationship, since other galaxies are almost absent
in the area of the Milky Way close to the Galactic
Equator. This, however, has nothing to do with their
real distribution in space. The enormous amounts of
gas and dust in our galaxy blocks the light of most of
the distant galaxies seen in that direction. So they
only appear to be absent. The galaxies are obviously
grouped into what astronomers call 'clusters' and
'superclusters', with dimensions beyond our
imagination.

In the bottom-right quadrant of the map you see
two large galaxies. These are the Large and the Small
Magallenic Clouds. The two clouds are regarded
as satellites of our own Milky Way, and therefore
are shown in blue on the star charts (14, 15 and 20).
However, since they are officially catalogued as
galaxies, they have been plotted on this map
as well.

The constellations

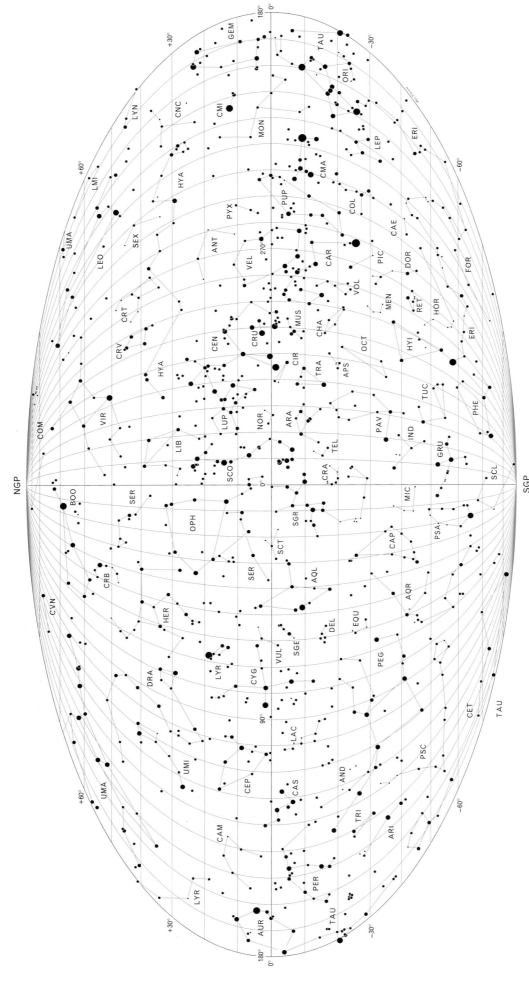

Mollweide's Equal Area Projection

Galactic Coordinates

Distribution of open clusters

Mollweide's Equal Area Projection

Galactic Coordinates

Distribution of globular clusters

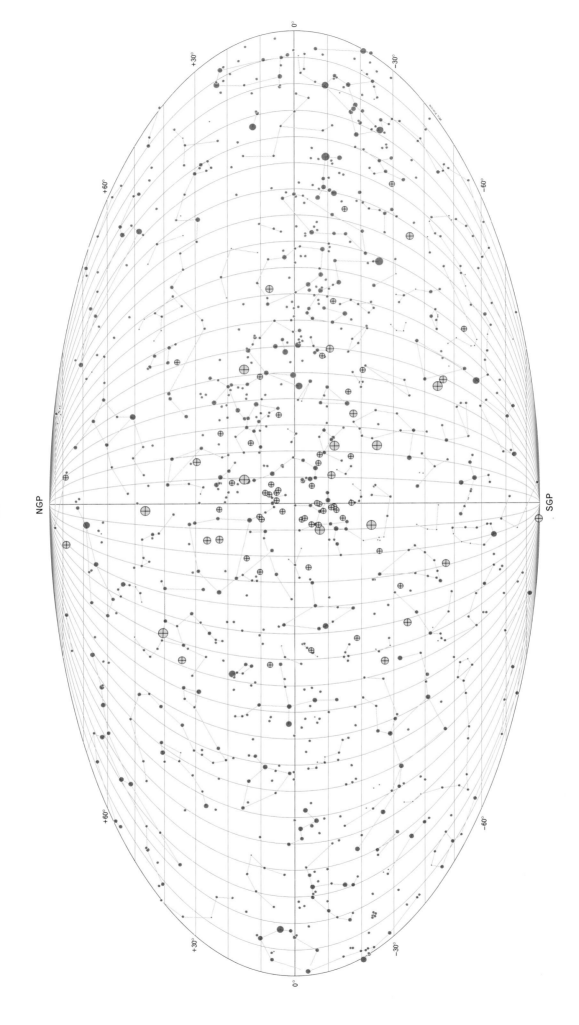

Mollweide's Equal Area Projection

Galactic Coordinates

Distribution of planetary nebulae

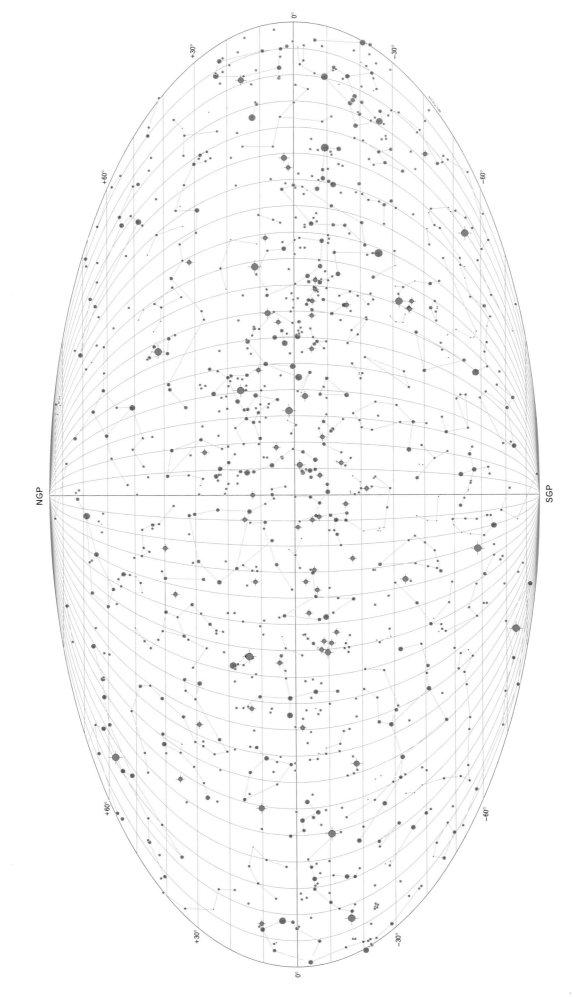

Mollweide's Equal Area Projection

Galactic Coordinates

Distribution of diffuse nebulae

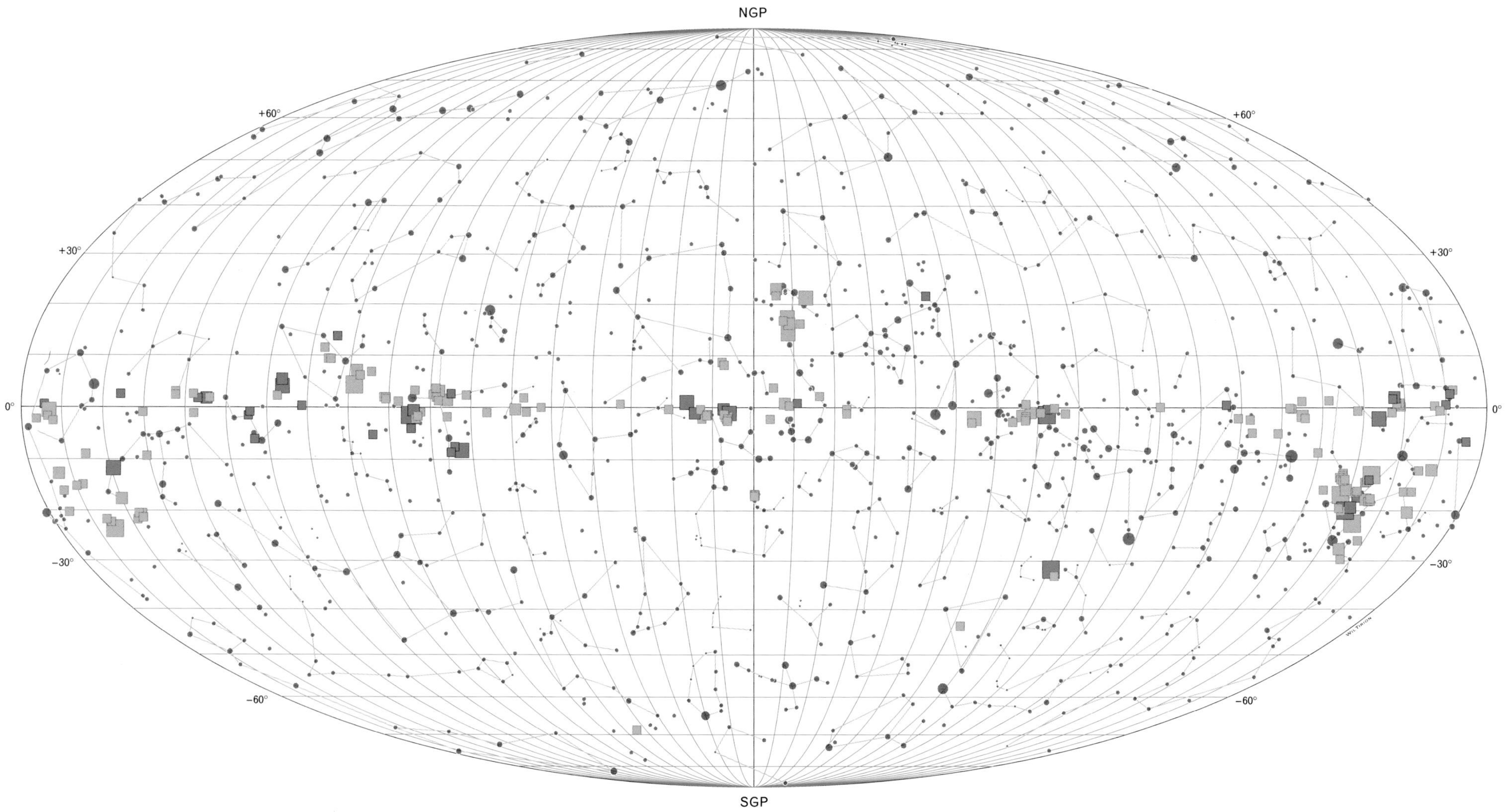

NGP

SGP

+60°

+30°

0°

−30°

−60°

+60°

+30°

0°

−30°

−60°

Mollweide's Equal Area Projection

Galactic Coordinates

Distribution of galaxies

NGP

SGP

+30°

+60°

−30°

−60°

Mollweide's Equal Area Projection

Galactic Coordinates

SOURCES AND REFERENCES

Catalogues

Dorrit Hoffleit, *The Bright Star Catalogue*, fourth edition, Yale University Observatory, New Haven, Connecticut, 1982

Smithsonian Astrophysical Observatory Star Catalog, Smithsonian Institution, Washington, 1966.

Alan Hirshfeld and Roger W. Sinnott, eds., *Sky Catalogue 2000.0*, Volume 1, Sky Publishing Corporation, Cambridge, Massachusetts and Cambridge University Press, Cambridge, England, 1982

Alan Hirshfeld and Roger W. Sinnott, eds., *Sky Catalogue 2000.0*, Volume 2, Sky Publishing Corporation, Cambridge, Massachusetts and Cambridge University Press, Cambridge, England 1985

Roger W. Sinnott, ed., *NGC 2000.0*, Sky Publishing Corporation, Cambridge, Massachusetts and Cambridge University Press Cambridge, England, 1988

Atlases

Wil Tirion, *Sky Atlas 2000.0*, Deluxe Edition, Sky Publishing Corporation, Cambridge, Massachusetts , 1981

Wil Tirion, Barry Rappaport and George Lovi, *Uranometria 2000.0*, Volume 1, Willmann-Bell Inc., Richmond, Virginia, 1987

Wil Tirion, Barry Rappaport and George Lovi, *Uranometria 2000.0*, Volume 2, Willmann-Bell Inc., Richmond, Virginia, 1988